The Forgotten Cure

Anna Kuchment

The Forgotten Cure

The Past and Future of Phage Therapy

Copernicus Books
An Imprint of Springer Science+Business Media

Anna Kuchment
Editor, Advances
Scientific American
New York, NY 10013, USA
akuchment@sciam.com

© Springer Science+Business Media, LLC 2012

Published in the United States by Copernicus Books,
an imprint of Springer Science+Business Media.

Copernicus Books
Springer Science+Business Media
233 Spring Street
New York, NY 10013
www.springer.com

Library of Congress Control Number:
2011940296

Manufactured in the United States of America.
Printed on acid-free paper

ISBN 978-1-4614-0250-3   e-ISBN 978-1-4614-0251-0
DOI 10.1007/978-1-4614-0251-0

For Mark, Eliza and my parents.

Acknowledgments

I'd like to thank, first and foremost, my sources: the scientists and entrepreneurs who trusted their life's work to me and sat by patiently as the book wound its way through the reporting, writing and publishing process.

I'm also grateful to my production editor at Copernicus Books, Arthur Smilios, to my agent Gary Morris, and to my first editor, Paul Farrell, who acquired the project. George Hackett, my former editor at Newsweek, carefully read and edited the final draft.

This book would never have come into being were it not for Fred Guterl, my science editor at Newsweek International, who first assigned me a story on phage therapy in 2001.

A large cast of former and current colleagues allowed me to take time off to work on this book over the years and also filled in for me while I was out. These include: Michael Elliott, Alexis Gelber, Fareed Zakaria, Steven Strasser, Nisid Hajari, Adam Piore, David Stone, Brian Connolly, and John Tucker.

Jerry Adler, Sam Freedman, Claudia Kalb, Shannon Jones, Allegra Wechsler, Anne Underwood, Karen Springen, and Will Dobson provided important encouragement along the way. Andy Nagorski was indispensable when it came to logistics and overseas reporting advice. Mark Ptashne took time to read a draft of the book and helped me address some of its early shortcomings.

For research help, the Archives of the Pasteur Institute, the Newsweek Research Center, the Los Angeles Times Library, and the Tom Mix Museum were incredibly generous and supportive. The Columbia University policy of allowing its alumni virtually free access to Butler Library is a boon that made researching and writing this book a true pleasure.

I may never have finished this book were it not for Ilyon Woo, who sets a high bar when it comes to talent, time management and determination. Her example, combined with her encouragement and outright nagging, kept me from giving up on the project.

My parents, who, like me, come from the part of the world where phages are still in use, were the inspiration for this book. My aunt and uncle, Sophie Vilker and Adriaan Jobse, asked often about its progress, and I deeply appreciated their support.

Hal and Jane Lamster took care of my daughter, Eliza, when book writing kept me away. This would simply not have gotten done without them. My extended family-by-marriage, Annie, David, Alex and Susan Stone, all provided encouragement, entertaining medical stories to keep me inspired, and crucial fact-checking help.

Above all, I'd like to thank my husband, Mark Lamster, for putting up with my erratic writing schedule and for taking on long stretches of single-parent time and especially for setting an example of what book writing success looks like. And thank you, Eliza, for always asking about my book and for excitedly reporting to your preschool teacher when it was done. You and your dad always made it a joy to set the book aside and come home.

Prologue

In April 2002, Fred Bledsoe was doing construction work on his parents' lake house near Fort Wayne, Indiana, when he stepped on a rusty nail. As sharp as it was strong, the nail bore through Bledsoe's shoe and lodged in the sole of his foot. He cleaned up the wound and drove to a nearby hospital where doctors gave him a tetanus shot. "Case closed", he thought.

One week later, his foot swelled up, and the wound began oozing pus. Bledsoe, who has diabetes, went to see his brother, Larry, an internist, who prescribed a 10-day course of antibiotics. That seemed to do the trick; Bledsoe felt much better. But in August, the infection came back again with a vengeance. He grew feverish, his foot tripled in size – "it looked like a football," said his sister, Saharra – and a new infection site appeared at the base of his big toe. He could barely get out of bed. Dr. Bledsoe had his brother hospitalized, and physicians started him on a heavy course of IV antibiotics. But diabetes had damaged the circulation in Fred's foot, making it more difficult for the antibiotics to penetrate deeply enough and for his body's own white blood cells to help beat back the invading organisms. Nine weeks later, Bledsoe's physicians gave up. "My doctor told me I'd have a good quality of life without my toes," Fred says.

In desperation he called Saharra, his closest confidant in a six-sibling family. Growing up in Fort Wayne's impoverished, predominantly black South Side, Saharra, now 50, took on the role of keeping her younger brother out of trouble. Now, 35 years later, she saw this as another chance to come to his aide. She was convinced that doctors weren't doing enough to save Fred's foot. "There had to be something, somewhere, that could be done," she thought.

The answer arrived a short time later via an episode of the CBS news program "48 Hours." Saharra was on her way out of the house, when the show came on and caught her attention. The segment was called "Silent Killers" and discussed the growing problem of antibiotic-resistant infections. After reporting on two scary incidents: a woman whose paper cut turned into a blood infection; an 18-month old girl who nearly died from an ear infection, the story turned to a case that was eerily similar to Fred's. Alfred Gertler, a jazz musician from Toronto, Canada, had developed an infection in his ankle after fracturing it while hiking in Costa Rica. As with Fred, doctors had advised him to have his foot amputated.

But Gertler refused to listen to his physicians. Desperate for an alternative, he had scoured the medical literature until he came upon a magazine article describing a treatment called "bacteriophage therapy." Practiced in the United States until the 1940s and still used in parts of Eastern Europe, it pits tiny viruses – bacteriophages, or phages for short – against disease-causing bacteria. These viruses, the most ubiquitous organisms on earth, are bacterial parasites: they reproduce by attacking and destroying deadly germs.

The world's oldest institute dedicated to the study and practice of phage therapy is in Tbilisi, the capital of the former Soviet Republic of Georgia. Unable to find a cure in the state-of-the-art hospitals of Canada, Gertler bought himself a plane ticket to the Third World. There, doctors infused his wound with an amber-colored broth teeming with the invisible creatures. After 3 days of treatment, he reported, his infection was gone. Though it recurred later, after he returned to Canada, it was in a mild enough form that doctors could finally fuse together his anklebones.

Saharra, whose father is a Baptist minister, saw the program as a sign from God. She called Fred and Larry and then spent the night surfing the Web, reading everything she could find on phage therapy. She contacted Betty Kutter, a phage biologist at Seattle's Evergreen University who maintains close ties with Tbilisi. Kutter warned Saharra that the process of treatment was lengthy and that it did not work for everyone. When Saharra insisted, she put her in touch with the microbiologists who had prepared Gertler's phages. They were from Tbilisi's G. Eliava Institute of Bacteriophages, Microbiology and Virology, established in the 1930s by the French-Canadian discoverer of bacteriophages, Felix d'Herelle and his close friend, the Georgian bacteriologist Georgi Eliava. The treatment, Saharra was told, could take as long as 30 days and would cost $2,000.

At first, Larry Bledsoe resisted the idea. "You know your sister, she'll try anything," he grumbled to Fred. But, prodded by Saharra, he did some of his own research. He learned that bacteriophages exist naturally in the soil, in tap water, in lakes and rivers, even in people's guts and nasal passages. After speaking with Kutter, Larry determined that the treatment, even if it didn't cure his brother, most likely would not hurt him. He gave his consent.

The last step was raising the money. The family, which lives in one of the poorest areas of Fort Wayne, pooled its resources to buy the plane tickets, and the Eliava Institute agreed to let them pay for the treatment in installments. Fred and Saharra packed their bags.

At 5 am, on an early November day, Saharra and Fred touched down at Tbilisi's small international airport, unsure of what to expect. Georgia, which had been independent from the Soviet Union for more than a decade, was at that time one of the poorest and most unstable of the former republics. Rent by a conflict over two breakaway provinces, Abkhazia and Ossetia, it faced a refugee crisis and mounting crime.

Zemphira Alavidze, a dignified woman in her sixties who had treated Gertler and who runs one of the oldest phage therapy labs at the Institute, met them with her husband and an English-speaking friend. Together they drove from the airport along unlit roads, headlights occasionally illuminating a dead dog or a street vendor sleeping beside his fruit stand. When Saharra spotted a curious looking billboard with the face of a middle-aged man staring out beneath Hebrew-like Georgian lettering, Alavidze explained that it was a missing person's announcement. Not long before the Bledsoes arrived, a British banker had been kidnapped from downtown Tbilisi in broad daylight. Fred turned to Saharra: "What have you gotten me into?"

The next day, Alavidze drove the Bledsoes to Republic Hospital, a large cinder-block structure with knocked-out windows and stray cats meowing in the small weedy yard outside. Inside, Alavidze showed them to an elevator that would take them upstairs. It was operated by an old man who made his living off the 10 tetri (about 2 cents) fee he charged per ride. Because the call buttons were broken, the shaft reverberated with the sound of people banging on the doors and yelling out their floor numbers in Georgian.

Fred was admitted to the hospital, where Saharra was allowed to share his room. Soon, their doctor, Chief of Surgery Gouram Gvasalia, arrived and explained that his hospital would attempt to heal Fred's whole body – not just the infection in his foot. His circulation was poor and his blood sugar was high, so they would put him on a diet and try to wean him off the massive doses of insulin he had been taking. Meanwhile, Alavidze would take a bacterial sample from his foot and test it against the phages in her lab to see which ones would work.

***

Under an electron microscope, bacteriophages look like insects from outer space. They have a round, polyhedral head, elongated bodies, a tail, and spindly, spider-like legs. Just one-fortieth the size of a bacterium, they eviscerate their prey in a meticulously choreographed operation: they start by clinging to the wall of a bacterial cell and, like a syringe, injecting their DNA inside. There, the DNA particles operate with stealthy efficiency, shutting down the cell's reproductive machinery and reprogramming it to make phages instead of bacteria. In the span of about 30 min, the phage produces hundreds of offspring inside its unwilling host, creating a brood of new "daughter phages" that burst from the cell, destroying it and scurrying off in search of more prey.

Unlike antibiotics, bacteriophages make more of themselves as they work, eventually outnumbering and eradicating the bacteria they were sent to destroy. But, while antibiotics are effective against a wide variety of bacteria, each phage is specific, meaning that microbiologists must spend days and sometimes weeks in the lab identifying the bacteria in a patient's tissue sample and finding a phage that will eradicate it.

The diagnostic center at the Eliava Institute determined that Fred's infection was caused by two types of bacteria, *Pseudomonas aeruginosa* and *Staphylococcus aureus*, and Alavidze and her coworkers set to work. They grew up bacteria from Fred's wound in a series of petrie dishes, each containing two cloudy stripes of a single germ. Alavidze keeps her pseudomonas phages, of which she has around 20, in small glass bottles capped with eye droppers. A coworker of hers placed one drop of the first phage on the left side of one of the pseudomonas stripes. The next phage she dripped over the opposite end of the stripe, and so on. Each bacterial stripe got two different phages, four per plate. Then, she put the plates in an incubator for 18 h, to allow the phages to reproduce and do their job. The next morning, they read the results.

Some phages did not work at all. In these areas, the stripe was as cloudy and opaque as it had been the day before. Others had left patchy circles – small areas where some of the bacteria had been eaten away. Only one phage had worked perfectly: where the coworker had dripped it, it had eaten away a clear circle where the foggy bacterial growth had been. This is the one they would use for Fred. They performed a similar experiment on Fred's strain of staph bacteria, and then mixed, multiplied, sterilized their phage solution and poured it into a set of small glass vials that were sealed shut over a Bunsen burner. The process took 10 days.

In the meantime, the Bledsoes were getting to know their neighbors at the hospital. Across from them lived a family of refugees from Abkhazia, one of Georgia's two breakaway provinces. The two simmering conflicts had uprooted 10% of the country's population, and there was not enough housing for all of them. The hospital had given the family a room where they installed a makeshift kitchen and made themselves at home. They frequently invited Fred and Saharra across the hall for lunch and dinner. The matriarch of the family had a small loom on which she wove handicrafts to sell on the streets. She made Fred and Saharra each a pair of socks and a small tapestry with an illustration of Georgia woven into it.

When Fred's phage preparation was ready, doctors doused his foot with it. Three times a day, a nurse would come, take two glass vials of phage out of a cardboard box, cut the tip of the vial off with a razor blade, transfer its contents into a syringe and squeeze it over Fred's toe. Physicians also put him on a low-sugar, low-fat diet and helped improve the circulation in his feet by administering electrical stimulation to the area. After 30 long days, his wound finally healed – and he had lost 19 lb on his diet. What was once a gaping hole that would not crust over had become a large but benign callus. Fred had arrived on crutches but left on his feet.

***

Bledsoe's case exposes deadly gaps in one of the world's most advanced medical systems. After penicillin was first mass-produced in the mid-1940s, wealthy nations enjoyed decades of relative peace of mind when it came to infectious diseases. Pharmaceutical companies pumped a steady stream of antibiotics into the marketplace, drugs that tamed once-fatal disease like pneumonia and strep throat. But, as patents expired and germs seemed to have been bowed into submission, that once fertile pipeline has dried up. Major drug companies have turned their attention toward newer, more profitable areas, like the diseases of aging: hypertension, heart disease, and diabetes. Patients take these drugs for life, while a course of antibiotics can last as little as a few days.

As a result, germs have made a comeback. So-called superbugs, bacteria that are resistant to one or more antibiotics, are on the rise across the United States. These bacteria used to be confined to hospital wards, but they are increasingly seeping out into the environment, where they infect otherwise healthy adults and children. From 1999 to 2008, the rate of children admitted to hospitals with methicillin-resistant *Staphylococcus aureus* (MRSA), one of the strains that infected Bledsoe, has grown tenfold[1] At the moment, there are still drugs to fight MRSA, but a growing number of bacteria are impervious to every antibiotic available. In a January, 2009 report, "Bad Bugs, No Drugs," the Infectious Diseases Society of America wrote, "There is an urgent, immediate need for new agents with activity against these panresistant organisms. There is no evidence that this need will be met in the foreseeable future." Just as in the pre-antibiotic era, doctors are battling to save the lives of patients with pneumonia, cuts and sinus infections.

Phage therapy holds potential as an important new weapon in the fight against superbugs. Rediscovered in the West in the mid-1990s, the treatment has brought a steady stream of venture capitalists, entrepreneurs and physicians through the Eliava Institute's halls. Independently, and with the help of specialists there, Western biotechnology companies are exploring ways of using phages to battle these deadly infections.

Once dismissed as a backward treatment, phage therapy has gained important ground in the last several years. In 2009, British company Biocontrol Limited completed the first double-blind clinical trials showing that phage therapy is safe and effective for the treatment of chronic, antibiotic-resistant ear infections. The United States Army is funding research into whether phages can heal some of the hardest to treat wound infections in Iraq war veterans. Meat and seafood companies are spraying the viruses on their equipment to protect consumers from foodborne illness. And researchers are exploring ways that phages can treat illnesses as diverse as lung infections in cystic fibrosis patients, breast infections in nursing mothers, sinusitis and chronic urological infections. In some ways, phages fit perfectly with the conventional wisdom that simple and natural products top artificial and chemically enhanced ones; one company has had its phages certified organic, Kosher and Halal.

But phages are no magic bullet. Critics point out that they can cause disease as well as cure it; by mingling their own genes with those of bacteria, phages have given rise to some of our worst killers, including diphtheria and food poisoning caused by *E. coli* 0.157. And, just like antibiotics, they breed resistance, though phage researchers say isolating a new phage is faster and cheaper than synthesizing a new antibiotic. Rapid genetic sequencing techniques help keep out so-called "lysogenic" phages that can pass dangerous genes to bacterial cells. While some still see phage therapy as a cultish phenomenon backed by weak science, the current crop of biotech startups is beginning to prove them wrong.

---

[1] Pediatrics "Antibiotic Management of Staphylococcus aureus infections in US Children's Hospitals, 1999 2008. Jason G. Newland et al. 2010; 125; e1294–e1300.

"The Forgotten Cure" weaves together the history of phages with the stories of scientists who've championed them at the risk of their careers and, occasionally, their lives. It will take you from the Pasteur Institute during World War I, through Stalin's Great Purge, to the Nobel podium and the bedsides of patients battling infections that no antibiotic can touch.

In the process, you'll learn that this treatment stands at the crossroads of two vastly different medical cultures. To the East: a country that provided free but substandard medical care. To the West: a country that offers superior medical care that not everyone can afford. Americans are accustomed to high-tech treatments and rapid-fire cures. Pharmaceutical and biotech companies, in order to keep the flow of innovation coming, expect vast profits.

Can the Western phage companies make money off an ancient, ubiquitous virus? Can they adapt this complex treatment to America's quick-fix culture? And, finally, can they solve the long-held medical mystery at the center of the story: are bacteriophages a long-forgotten cure for deadly, reemerging infections or an unreliable folk medicine with the false gleam of fool's gold?

Contents

Chapter 1
Helpful Little Bodies

As dusk settled across Los Angeles one November evening in 1931, an American screen idol lay near death at Hollywood Hospital. Tom Mix, the first great Western film star, famous for his gravity-defying riding stunts and diamond-studded spurs, had battled armed train robbers, charging Indians and wild broncos. Now, a sudden and severe stomach infection brought on by a ruptured appendix was proving to be a far more daunting foe.

Mix, 51, was brawny and handsome, with thick black hair and a strong, cleft chin. He and his brown-and-white horse, Tony, had risen to fame on the silent films of the prior 2 decades, known for their daring acrobatics and flamboyant attire: a ten-gallon Stetson and white tailored suits for Mix, elaborately carved saddles for Tony. But the arrival of sound in the late 1920s pushed them off the silver screen. After a short hiatus from film, during which Mix traveled with a popular circus, the actor was on the verge of a comeback. Universal Pictures had just signed him to a multi-movie deal that would allow audiences to hear the actor's voice for the first time.

Just before the cameras were set to roll, Mix fell ill. He was home at his Hollywood mansion when he developed a stomachache.[1] At first he didn't think anything was seriously wrong. But that night the pain grew worse and, at 10 p.m., he had his physician, Dr. R. Nichol Smith, called to the house, which was easily identifiable by the initials "T.M." boldly displayed outside in lights. Smith diagnosed appendicitis and had the actor rushed to the hospital for emergency surgery.

By the time doctors took a scalpel to Mix, his appendix had ruptured, dumping millions of bacterial cells into his peritoneum, the abdominal cavity that stretches from the bottom of the diaphragm to the base of the pelvis. Peritonitis, an often fatal infection that attacks the lining of this cavity, had set in. Mix was in unrelenting pain, unable to sleep or eat solid food. "His iron constitution will go a long way in determining the final outcome," said Dr. Smith. "And he is putting

---

[1] Paul E. Mix, *The Life and Legend of Tom Mix* (New York: A. S. Barnes and Company, 1972), 101–103.

A. Kuchment, *The Forgotten Cure: The Past and Future of Phage Therapy*, DOI 10.1007/978-1-4614-0251-0_1, © Springer Science+Business Media, LLC 2012

up a courageous fight."[2] Fan mail immediately poured into the hospital at a rate of 50 letters an hour, and the case was tracked in major newspapers across the country.[3] "It sure will be rough on a lot of western folks if Tom goes over the Big Divide," said Pat Chrisman, a close friend of Mix and a popular cowboy actor of the day.[4] If the defenses of Mix's own body, or a drug, didn't stop the infection in time, it would spread into his bloodstream and begin shutting down his organs, one by one.

In Tom Mix's day, the prognosis for peritonitis was "very bad," according to one of the most widely used medical reference book of the time.[5] Screen heartthrob Rudolph Valentino had recently died a highly publicized death from the same illness, and the official fatality rate hovered somewhere between 40 and 70%.[6] Antibiotics would not come into general use until the introduction of penicillin in the mid-1940s, and Mix's physicians had few tools with which to fight his infection.

The standard treatment for peritonitis was to cut open the patient's stomach and rinse the abdominal cavity with a salt solution, which was little more effective than doing nothing at all.[7] Sometimes, doctors would also give intravenous injections of antibodies, then called "antitoxin" or "immune serum," that had been harvested from horses or sheep. Made famous by Louis Pasteur's protégé Emile Roux, who pioneered their use in 1891 against diphtheria, serums had been developed for a variety of infectious diseases, including mumps, bubonic plague and pneumonia.[8] They were the standard treatment of the day but had their limitations. They could be used only against a small number of illnesses (cholera, for instance, remained impervious),[9] delivered mixed results and sometimes came with harsh side-effects like rash, fever and anaphylactic shock.[10] Overall, science was not yet winning the battle against microbes, leaving doctors strained beneath enormous caseloads of fatal sore throats and coughs, often unable to offer much beyond sympathy. "Disease was looked upon as the striking arm of Providence, the punishment of sin and immorality," wrote a physician in 1949, looking back on the pre-antibiotic era. "Even when, with the progress of science, we acquired full knowledge of the disease-producing germs, the fatalistic attitude toward infectious disease still remained.

---

[2] "Tom Mix in Desperate Life Fight," *Los Angeles Times*, 25 Nov. 1931, sec. 2, p .1.

[3] "Cowboy Film Idol Out of Shadow," Los Angeles Times, 3 Dec. 1931, sec. 2, p. 1.

[4] Robert S. Birchard, *Tom Mix and the Movies* (Burbank: Riverwood Press, 1993), 232.

[5] Modern Medicine: Its Theory and Practice, Sir William Osler, ed. Vol. III: Diseases of Metabolism – Diseases of the Digestive System, pg. 928 Lea & Febiger, Philadelphia and New York, 1926.

[6] 40 percent figure from Osler's. 70 percent from M. Kirschner, "Die Behandlund der akuten eitrigen freien Bauchfellentzundung. *Langenb Arch Chir* 142 (1926): 253–267, cited in Thomas Genuit, M.D., "Peritonitis and Abdominal Sepsis," *eMedicine.com* (2002).

[7] Author interview with Sidney Raffel, August 2003.

[8] A Text-Book of Medicine, Russell L. Cecil, ed. 1930, W. B. Saunders Company, Philadelphia and London, pgs. 38, 267, and 299.

[9] Cecil, 272.

[10] Osler's pg. 741.

Although modified in some degree, this belief was still part of our consciousness until the time of the discovery of penicillin."[11]

But there was one new bright spot in this battle against infection: a tiny organism discovered independently by scientists working on opposite sides of the English Channel, one in 1915, the second 2 years later, in 1917. Although microscopes were not yet powerful enough for Frederick Twort, a British bacteriologist, and Felix d'Herelle, an independent-minded French-Canadian researcher working at the Pasteur Institute in Paris, to glimpse their discoveries, the agent would turn out to be a virus that preyed on bacteria 40-times its own size. D'Herelle named the organism "bacteriophage," meaning "developing at the expense of" bacteria,[12] or "bacteria eater." For Twort it was a passing observation, but for d'Herelle it would become a lifelong occupation.

By the early 1930s, phage therapy was enjoying the pinnacle of its use in medicine, though it was still considered an experimental treatment. The press hailed phages variously as "nature's G-men,"[13] "infinitesimal friend[s] of mankind,"[14] and "helpful little bodies."[15] They had been popularized in Sinclair Lewis's 1925 novel Arrowsmith, in which an ambitious young scientist uses phages to treat victims of plague in the Caribbean. Lewis, in 1930, became the first American writer to be awarded the Nobel Prize in literature, partly on the basis of this work. Major pharmaceutical companies, like Eli Lilly and E. R. Squibb,[16] marketed phage preparations, and physicians in the United States, Europe and Asia administered them for ailments ranging from urinary tract infections to cholera and dysentery. The results were mixed, but when it did work it brought such rapid and dramatic results as to seem nearly miraculous.

Not all physicians were aware of phage therapy. Mix was fortunate that Smith, a prominent celebrity doctor who would return to the headlines a decade later after rescuing Babe Ruth from a bout of pneumonia, made a point of reading the latest medical journals and carefully testing new remedies. He knew, for example, that E. W. Schultz, a professor at Stanford University's department of bacteriology and experimental pathology had recently established a phage lab. For about $2 – the cost of expenses – Grace Shields, a friendly middle-aged technician and the lab's lone

[11] B. Sokoloff, *The Miracle Drugs* (Chicago: Ziff-Davis, 1949), 253, quoted in Karen Ho, "Bacteriophage Therapy for Bacterial Infections: Rekindling a Memory From the Pre-Antibiotics Era," *Perspectives in Biology and Medicine* 44, no. 1 (2001): 1.

[12] William C. Summers, *Felix d'Herelle and the Origins of Molecular Biology* (New Haven: Yale University Press, 1999), 192, no. 2.

[13] G.A. Skinner. 1937. Nature's G-Men. *Hygeia* 15 (March): 243, cited in Karen Ho.

[14] Edith L. Weart, "A New Foe of Germs," *North American Review*, July 1929, 33.

[15] "Journal aids study of bacteria-gobbling phage," *Newsweek*, Dec. 15, 1934, 23, cited in Ho, Karen, 6.

[16] Margaret E. Straub and Martha Applebaum, B.A., "Studies on Commercial Bacteriophage Products," *Journal of the American Medical Association* 100, no. 2 (1933): 110.

staff person, would prepare phages for any physician who requested them.[17] Smith felt that Stanford provided the "only reliable preparation" available on the West Coast.[18] "The 'phages obtained from commercial pharmaceutical houses have proved to be inert by the time they reach their destination," he would say in May 1932 before a joint meeting of the Los Angeles County Medical Association and the Los Angeles Surgical Society.[19] The doctor immediately ordered that a shipment of Schultz's phages be sent by plane to Los Angeles and, after Mix's surgery, Smith sent a sample of his patient's bacteria to Stanford so that Shields could make an exact match.

At 6 p.m. on Tuesday Nov. 24, just 1 day after Mix had been hospitalized, a small propeller plane touched down outside Glendale Airport's Spanish-style terminal building carrying a supply of phages from Stanford that workers promptly loaded onto a waiting motorcycle and rushed to Hollywood Hospital. There, doctors administered the preparation to Mix. Though newspaper accounts claimed the phages were introduced into Mix's bloodstream through an IV, Smith would later say that, for cases of peritonitis, he preferred to inject the bacteriophage directly into the belly, via a catheter. Either way, Smith remained cautious about his patient's prognosis. "I do not want to appear gloomy about Mr. Mix's condition," he told reporters on Wednesday. "But he is in a precarious state and only time will tell."[20]

Over the course of 4 days, at least three shipments of phages were sent to Hollywood Hospital[21] and Mix's doctors believed they quickly began to work. "Tom Mix Rallying Slowly," reported The New York Times on Saturday. Two days later, Dr. Smith said the actor was "brightening up." On Dec. 3, the Los Angeles Times ran a large, front-page photo of a bonnet-clad, smiling nurse, feeding Mix his first solid food in 8 days. "Rounding Up Grub," ran the caption. "It was the closest I had ever come to death," Mix told the newspaper. Just 5 months after Mix recovered, "Destry Rides Again" appeared in theaters. The actor went on to make eight more talkies.

The Father of Bacteriophages

At the time that Tom Mix fell ill, Felix D'Herelle, who had discovered phages while working at France's Pasteur Institute in 1917, was working on the opposite coast of the United States. He had been appointed professor of bacteriology at the Yale School of Medicine, a position that would mark the peak of d'Herelle's

[17] Author interview with Sidney Raffel, Aug. 2003.

[18] R. Nichol Smith, "Advanced Treatment in Postoperative Ileus," *American Journal of Surgery* 19, no. 2 (1933) 273.

[19] Ibid., 272.

[20] "Tom Mix Fights Against Death," *Los Angeles Times*, 26 Nov. 1931.

[21] "Tom Mix Rallying Slowly," *New York Times*, 28 Nov. 1931. (Accessed electronically).

academic career. The Montreal-born, Paris-raised scientist, who sported a handlebar mustache and a pointy goatee, was an adventurer with a prickly personality. He was self-confident, idealistic, outspoken and unafraid of controversy – qualities that helped him popularize bacteriophage therapy around the world but prevented him from winning the lasting respect and acceptance of the scientific establishment. Unlike Sinclair Lewis, he never won the Nobel Prize. "His experimental and technical talents far exceeded his abilities as a scientific negotiator in the complex social world of international science," writes his biographer, William Summers.[22]

D'Herelle developed a passion for microbiology at a young age. Born in Montreal in 1873 into a privileged family, he moved to Paris with his mother after the death of his father in the late 1870s. Upon graduating from lycee, he received a gift from his mother of 3,000 francs, enough to travel through South America for 3 months. As he was returning to Paris from Rio de Janeiro by ship, yellow fever broke out among the passengers and crew. Within 8 days, 20 people were dead, and d'Herelle watched as their bodies were floated out to sea, one by one. At that moment, because he did not panic, d'Herelle saw how well suited he might be to the field of infectious diseases. "It is probable that I have, by birth, the first required quality needed to make a good microbe hunter," he wrote in his memoirs. "Most of the passengers were in anguish: I was perfectly calm, I thought I was invincible."

---

[22] Summers, vii.

Having never formally attended college, d'Herelle was entirely self-taught in the sciences. In 1894, he moved back to Montreal with his new wife, Marie – he was 21, she was all of 16 – and set up a home laboratory. At the time, the study of bacteria was a nascent field, and d'Herelle would become one of its pioneers. He recalled in his memoirs, "I was always thinking about bacteriology, so on my arrival [in Montreal] I set up a laboratory and began to experiment, all alone because at this time there were only two French Canadians who were interested in microbes, Dr. Bernier, who was later the first professor in this subject at the University of Montreal, and myself."[23] D'Herelle made up for his lack of a formal education by plunging directly into field work. Following in the footsteps of his hero Louis Pasteur (1822–1895), d'Herelle accepted a government appointment, earned through a family connection, to study fermentations. (It was Pasteur who had discovered that microorganisms are responsible for turning sugar into alcohol.) In this case, the task was to turn excess Canadian maple syrup into whisky. One appointment led to another, and soon d'Herelle was traveling through Mexico and Guatemala, Argentina and Algeria, working on fermentation and pest control – including an innovative way of exterminating locusts by infecting them with bacteria.

In 1911, D'Herelle and his family took up residence in Paris where he found work as an unpaid assistant at the world-famous research center dedicated to his idol, the Pasteur Institute. Though d'Herelle wouldn't be earning a salary – at least at first – his compensation would come in the form of access to some of the most brilliant minds in medical research. Pasteur had proved that infectious diseases are caused by living organisms and had developed the first vaccine against rabies. The mission of the Institute was to spread its founder's knowledge and to build upon it, expanding rabies vaccination and probing the mysteries of how germs cause disease. Toward that end, it had established a network of affiliated institutes throughout France's colonial empire, stretching from Indochina through Northern Africa, where Pasteurians treated local populations and studied exotic illnesses. Meanwhile, in Paris, a group of dedicated investigators continued to make historic breakthroughs. It was here that Emile Roux, together with Alexander Yersin, in 1888 figured out why it took so few diphtheria germs to kill a patient (they emit a deadly toxin that spreads throughout the body), opening the door for the first reliable diphtheria treatment. In 1889, Roux inaugurated the world's first microbiology course ever taught at the Institute. Thanks to such efforts, observed one medical writer of the time, the public had reason to hope "that microbes were going to be turned from assassins into harmless little pets."[24] Within a few years, d'Herelle would add his own achievement to this prestigious roster. It would change the course of his career, revolutionize the treatment of bacterial infections in the 1920s and '30s and lay the groundwork for the development of microbiology's successor as the cutting-edge science of the day: molecular biology.

---

[23] Felix d'Herelle, Les peregrinations [sic] du'n microbiologiste, quoted in Summers, *Felix d'Herelle*, 5.

[24] Paul de Kruif, Microbe Hunters, pg. 178, Harcourt, San Diego, New York, London 1996. Originally published in 1926.

As with many great scientific achievements, d'Herelle discovered bacteriophages by chance. While working with sick locusts, he observed a puzzling phenomenon: amid some of his bacterial cultures, there were what he described as "*taches vierges*" – pure or clear spots on an otherwise cloudy background. When scientists wish to study the behavior of a particular microbe – whether for diagnostic or research purposes – one of their first steps is to grow the bacteria in a glass Petri dish coated with a clear, jellylike medium called agar. Under normal conditions, when the dish is placed for several hours in an oven heated to 37°C – the temperature of the human body and the optimal one for bacteria to replicate – bacterial cells begin to multiply and form cloudy patterns on the agar that resemble frost on a windowpane. But here, small, clear and perfectly circular spots had formed amid the bacterial growth.[25]

D'Herelle was intrigued by the phenomenon but hadn't known what to make of it. He had assumed the spots were somehow connected with the locusts' disease, yet had been unable to systematically reproduce it or to identify, under a microscope, anything that might be causing it. (The electron microscope would not be developed until 1940, and phages were too small to be seen through contemporary devices.) D'Herelle mentally filed away the observation and didn't think about it for several more years. While he was at the Pasteur Institute, World War I started, and its mission of monitoring disease outbreaks and manufacturing commercial vaccines took on new urgency.[26] In 1916, a severe epidemic of dysentery broke out among cavalry troops stationed at Maisons-Laffitte, on the outskirts of Paris and just 50 miles from the French-German front.[27] D'Herelle was asked to look into the outbreak – to analyze stool samples from the troops to see if they were suffering from a new, more virulent form of the infection. It turned out that they were. But, more important, it was while analyzing samples from these and other patients that d'Herelle saw his *taches vierges* once again. "I finally had proof," he wrote, "that the phenomenon of the clear spots wasn't limited to the coccobacilli [rod-shaped bacteria] of locusts, but that bacteria pathogenic to man were equally susceptible to it." Soon, d'Herelle noticed something even more striking: the *taches vierges* seemed to appear only in bacterial samples from patients who were recuperating. What he had at first believed to be an "instrument of malady," he thought, might in fact be an "instrument of recovery."[28] He immediately began to check his hypothesis.

A new dysentery patient had just been admitted to the hospital at the Pasteur Institute. D'Herelle decided that he would carefully monitor her progress, not only during the course of her illness but also throughout her recovery to see at which point the clear spots would appear. Each day, he took a sample of her stool, mixed a few drops of it with a few cubic centimeters (c.c.'s) of meat bouillon, a clear yellow liquid on which bacteria feed and multiply, and passed the mixture through a ceramic

[25] d'Herelle, Felix. "Perigrinations d'un Microbiologiste," unpublished manuscript at the Archives of the Pasteur Institute, Paris. p. 256.

[26] Summers.

[27] Summers, *"Felix d'Herelle."* 47.

[28] Felix d'Herelle, Perigrinations, 279, Archives de Inst. Past.

filter to sterilize it. The filter would keep out the bacteria but allow the much smaller source of the *taches vierges,* should it be present, to pass through. D'Herelle then added the filtered solution – the filtrate – to a test tube containing bacterial cells for dysentery. There would be two ways to check for the presence of the agent. One would be to place the mixture into an incubator overnight to see whether or not it turned cloudy: a cloudy solution would indicate the presence of a dense growth of bacteria, a clear solution would mean that d'Herelle's invisible agent had gobbled up the bacteria, leaving just the clear broth and some bacterial debris. The other way would be to pour the solution over an agar coated Petri dish, incubate it, and see whether clear spots formed among the bacterial colonies. D'Herelle opted for the first method, the test-tube method. Each night, he would seed one test tube with a mixture of bacterial cells and filtrate and a second, "control" test tube with just the bacterial sample from the patient.

For the first 3 days of the patient's illness, nothing happened. The two test tubes, left in an incubator overnight, grew equally cloudy with bacterial growth. But then, wrote d'Herelle, "On the fourth day, in the morning, a very strange spectacle awaited me." The control test tube, filled with nothing but dysentery bacteria, was opaque as usual. But the second tube, to which he had added the filtrate, was as clear as it had been the night before. The dysentery bacteria had dissolved overnight "like sugar in water." "When I close my eyes," wrote d'Herelle, "I can still see that scene, which marked a poignant moment in my existence: holding the two tubes, one turbid, the other as limpid as unseeded broth, and that limpid tube was, for me, something wondrous. It was one of those rare moments of intense, absolute joy that a lucky researcher experiences two or three times during the course of his life and that largely makes up for all the sleepless nights, all the years of work."

D'Herelle rushed to the hospital to see if the patient was feeling any better. He ran into her nurse in the hallway.

"You will find a marked change," she told him, "I have never seen such a grave case of dysentery clear up so quickly."

D'Herelle was fortunate, he noted, to have come upon a patient who would not only recover, but in whom the healing process was so dramatic.

Challenge and Controversy

D'Herelle published his first paper on bacteriophages in 1917, in which he reported: "… I have isolated an invisible microbe with the properties of antagonism to the bacillus of Shiga,"[29] a strain of bacteria that causes a severe form of dysentery. It was a landmark finding not only because it heralded a promising new therapy, but because until then, bacteria were believed to be the smallest living things. He went on to explain the "very simple" process by which he had isolated it and to hypothesize that there were phages for many other types of diseases – indeed, he

---

[29] Felix d'Herelle. "Sur un microbe invisible antagoniste des bacilles dysenteriques," *Comptes rendus Acad. Sci. Paris* 165 (1917): 373, cited and translated in Summers, 185.

had already isolated one for typhoid. Most importantly, he laid out the reasoning that led him to conclude that phages were living organisms, a claim that many of his contemporaries would vociferously challenge. Displaying clever and meticulous lab work, he proved that phages were capable of reproducing within a culture of bacteria. He did this by diluting his phage mixture and showing that a small amount is just as effective as a large one. "The invisible microbe grows in the … culture of Shiga bacillus because a trace of this liquid, placed in a new culture of Shiga, reproduces the same phenomenon with the same intensity," he wrote.

But this paper, like much of d'Herelle's work, displayed a combination of dead-on instinct and oddly misplaced beliefs – perhaps not surprising, given that he was self-taught. While he immediately and correctly recognized that each phage is specific and can attack only one single strain or sub-species of bacteria, he also set forth the notion that phages could be adapted to destroy any microbe. "[The phage's] parasitism is strictly specific," he wrote, "but if it is limited to one species at a given moment, it may develop antagonism in turn against diverse germs by accustomization." He had reached his conclusion after successfully breeding his phages to be effective against other strains of dysentery. In fact, d'Herelle was partly right: because phages mutate rapidly, they can be "adapted" slightly to attack close cousins of a particular germ. But few if any phages are capable of attacking multiple species or strains of bacteria. Scientists now know that there are about as many different types of phages as there are types of bacteria: hundreds of thousands, if not millions, only a fraction of which have been identified.

The other conclusion that later proved incorrect was d'Herelle's supposition that a bacteriophage is the "true microbe of immunity" and that it is the agent primarily responsible for helping patients fight off disease. That posed a direct challenge to the work of d'Herelle's colleague Jules Bordet, director of the Pasteur Institute's Brussels lab, who in 1919 would win the Nobel Prize for identifying factors in the blood ("antibodies" and "complements") that destroy bacteria. In his 1930 book, "The Bacteriophage and Its Clinical Applications," d'Herelle wrote boldly of how his discovery had "crumbled" biologists' earlier theories on immunity. "The antibodies play no part in the phenomena of recovery," he wrote, a statement that has turned out to be false. Today, phages are believed to play little if any natural role in the healing process. Their function is ecological: keeping the bacterial population in check, much the way frogs and birds of prey prevent the earth from being overrun with insects and rodents.

Without television or radio, neither of which had yet been developed, word of d'Herelle's discovery spread slowly. The year 1920 brought the first substantial batch of papers responding to the phenomenon, including 11 by d'Herelle's Pasteur Institute colleagues – a substantial number, given that the institute and the country were still recovering from the war and the mobilized staff was just returning.[30]

While many scientists tried, and succeeded, in replicating d'Herelle's results in the lab, others disputed his findings. The most prominent of the naysayers was Bordet, who, in two papers published with coworker Mihai Ciuca in 1920,

---

[30] Summers, pg. 60–61.

portrayed d'Herelle's work within the context of his own theories.[31, 32] The two authors argued against the existence of bacteriophages. The bacterial cells d'Herelle had observed were destroyed not by phages, they wrote, but by an immune response on the part of the host animal. An animal's immune system triggered the bacterial cells to produce a lytic agent – an agent capable of destroying bacteria – leading to the cells' imminent death.[33]

Bordet and Ciuca did not end their criticism with these two papers. One year later, they announced that d'Herelle hadn't been the first scientist to observe this lytic phenomenon. "In the interest of historical accuracy," they wrote in a paper submitted at a meeting of the Belgian Society of Biology in 1921, "we are calling attention to a prior work which d'Herelle did not know of, and which we, ourselves, have ignored, that in truth, contains all the findings which d'Herelle has reported."[34] The paper, published in the British journal *Lancet* in 1915 – 2 years prior to d'Herelle's – was by Frederick William Twort, a bacteriologist who was superintendent of the Brown Animal Sanatory Institution in London.[35] Twort had observed that some of the bacterial colonies he had grown in his lab "often showed watery-looking areas ... [that] became glassy and transparent"[36] with time. Like Bordet and Ciuca, Twort surmised that the "active transparent material" was produced by the bacteria themselves.

D'Herelle, exhibiting the stubborn pride he was known for, at first insisted that the two scientists had witnessed different things. But he eventually put the issue of discovery aside, as did the rest of the scientific community, and accepted that phage may have had two independent discoverers. Thus, it became known for a time as the "Twort-d'Herelle phenomenon," a term that soon fell out of use to be replaced by d'Herelle's "bacteriophage."

Healing the Sick

The ongoing controversies on the nature of phage would last into the early 1940s, but they did not shake d'Herelle's confidence, nor did they dissuade him from putting his discovery into practice. Seeing the potential that bacteriophages held as a

---

[31] Bordet, Jules, and Ciuca, Mihai. 1920. "Exsudats leucocytaires et autolyse microbienne transmissible." *Comptes rendus Soc. Boil.* Paris 83: 1293–1295, as cited and analyzed in Summers, pgs. 65–66.

[32] Bordet, Jules, and Cuica, Mihai. 1920. "Le bacteriophage de d'Herelle, sa production et son interpretation." *Comptes rendus Soc. Boil.* Paris 83: 1296–1298, as cited and analyzed in Summers, pgs. 65–66.

[33] Summers, pg. 65.

[34] Jules Bordet and Mihai Ciuca. 1921. "Remarques sur l'historique de recherches concernant la lyse microbienne transmissible." *Comptes rendus Soc. Boil. Paris* 84: 745–747.

[35] Summers, Felix d'Herelle, 70.

[36] Twort, Frederick W. 1915. "An investigation on the nature of ultramicroscopic viruses." Lancet 2: 1241–1243, cited in Summers, p. 71.

weapon against infectious disease, he quickly set to work testing them in animals and humans. In people, he started with dysentery, since he already had the necessary phages. The disease, which leads to a high fever and bloody diarrhea, was a particular threat to children, the elderly and to troops, whom it infected through contaminated latrines.[37] To find his first patients, d'Herelle called upon the famous French pediatrician Victor-Henri Hutinel of the Hopital des Enfants-Malades in Paris. "He agreed to let me treat the young patients on the condition that I first demonstrate that ingestion of the cultures of bacteriophage were harmless," d'Herelle recalled in his memoirs.[38] In those days, before the advent of regulatory agencies that would demand double-blind studies, it was common practice for scientists to test medications on themselves before administering them to patients. D'Herelle, who had already ingested numerous ampoules of phage, proposed another trial on himself of a dose 100 times greater than the one he would give the children. Several hospital interns who were present also requested a "small glass" of phage to drink. On 1 August 1919 a large flask was "shared all around," and even Professor Hutinel tasted the culture: "Opinion was unanimous, the flavor, if not delicious, was not too disagreeable."[39] Soon, three brothers, aged three, seven, and twelve, whose sister had just died of dysentery, were admitted to the hospital in grave condition. Each patient was injected with phage and all began their recoveries within 24 h.

From there, d'Herelle turned his attention to testing phages against other bacterial killers of the day, including plague and cholera. He had quickly learned that bacteriophages are found wherever bacteria thrive: in sewers, in rivers that catch waste runoff from pipes, and in the stool of convalescent patients. Like any predator, bacteriophages are best able to survive and multiply when they are in close proximity to their food supply, where they fulfill their evolutionary role of keeping bacteria in check. To this day, phage workers tap those same sources to find new strains of the virus. D'Herelle isolated his first bacteriophage against plague in 1920, while working at the Pasteur Institute in Saigon, Indochina. It was the "land of [his] dreams," he wrote, because one encountered so many exotic diseases there that were worthy of study.[40] Unable to find any convalescent patients – the death rate from plague topped 60% – he obtained the phage from the feces of rats, which served as plague carriers.

His first chance to test these new phages came in 1925, while working in Alexandria, Egypt. D'Herelle had left the Pasteur Institute in Paris shortly after returning from Indochina, the result of a falling-out with its new director, Albert Calmette. He had accepted a new position as a bacteriologist for the Alexandria Quarantine Service. There, in the summer of 1925, four passengers aboard two ships docked at Alexandria's harbor were diagnosed with the dread disease.

[37] "Modern Medicine: Its Theory and Practice," Sir William Osler, ed. (Lea & Febiger, New York: 1925), 687.

[38] Felix d'Herelle, *Perigrinations*, 404, cited in Summers, *Felix d'Herelle*, 114.

[39] Felix d'Herelle, *Perigrinations*, 405, cited in Summers, *Felix d'Herelle*, 114.

[40] Felix d'Herelle, Perigrinations, 409, cited in Summers, Felix d'Herelle, 115.

Bubonic plague, transmitted by the bite of a flea that preys on both humans and rats, carried symptoms of a high fever with chills and dramatically swollen lymph nodes, called bubos, found most commonly in the groin. In his book, "The Bacteriophage and Its Behavior," d'Herelle described the case of 18-year-old Georges Cap.

> On July 13, at 2 p.m., the patient had a temperature of 40.3°C, the pulse was 130, the face was congested, the eyes were injected and drooping ... The two buboes were the size of nuts, and painful when pressed. I gave an injection of 0.5 cc. of Pestis-bacteriophage into each of the two buboes, the needle being introduced about the center of the bubo. Apparently the injection caused no pain, since the patient showed no reactions of defense ...
>
> On the morning of the 14th the condition of the patient was completely changed. He stated that he felt weak, but that he was not in pain and felt well. ...
>
> On the 15th he sat up in bed; and on the 16th he begged for food...
>
> By Aug. 8 healing was complete.[41]

D'Herelle's results, which were first published as the lead article in *La Presse Medicale* on October 21, 1925, were greeted with enthusiasm by the colonial government. Dr. A. Morison, the British representative on Egypt's Conseil Sanitaire, Maritime et Quarantenaire d'Egypte, which controlled d'Herelle's lab, wrote to C. E. Heathcote-Smith, the British Consul General in Alexandria about the report. "To my mind the article is very interesting and of intense importance, offering (as it appears to me to do) an almost certain cure for bubonic plague ... Dr. d'Herelle has supplied already the Sanitary Administration of Egypt with the necessary bacteriophage. I think India ought to arm itself."[42, 43]

As soon as word of d'Herelle's success reached India, Bombay's Haffkine Institute requested a shipment of his phage. The founder of this institute, the Odessa-born bacteriologist Vladimir Haffkine, had developed vaccines against cholera and plague, but they worked only prophylactically and could not cure patients who were already ill. In 1926, the institute performed a controlled trial of phage therapy in Hyderabad and Agra, in which alternate cases of plague were treated with bacteriophage. It failed, apparently because the viruses isolated in Indochina didn't work against the strain of plague bacteria active in Bombay. The following year, d'Herelle was back in India to organize a trial of phage therapy for cholera in the villages of the Punjab. Here, the results were more promising. In one group of 33 patients treated by conventional methods – injection of fluids and salts – the mortality rate was 40%, while in a separate group of 16 patients all treated with phage, none died. In another trial, d'Herelle collected information on 198 patients over 6 months. Of this group, 74 patients had received phage, and their mortality rate was 8%. In the untreated group, the mortality rate was 63%.

---

[41] Felix d'Herelle. 1926. "The bacteriophage and its behavior." Baltimore, Williams and Wilkins. 567–576.

[42] Summers, 126.

[43] A. Morrison to C. E. Heathcote-Smith, 4 December 1923, pp 118-119, L/E/7 #1425, File 7616, India Office Records, The British Museum, London, cited in Summers, *Felix d'Herelle*, pg. 126.

Once again, the results were met with enthusiasm, and further trials were ordered up. "A new light has appeared on the horizon in D'Herelle ... whose bacteriophage appears likely to revolutionize ... treatment of cholera," wrote Lt. Col. C. A. Gill, former director of public health in Punjab, to Sir Leonard Rogers, a cholera expert.[44] At this time, d'Herelle had begun anticipating his Yale appointment and turned down further offers to work in India. The cholera trials were carried on by Lt. Col. J. Morison, acting director of the Haffkine Institute and a convert to phage therapy,[45] and by Igor Asheshov, a Yugoslavian bacteriologist whom d'Herelle had recommended take his place. Their results, however, were inconclusive, due in part to poor funding and little cooperation on the part of the local Indian government and overworked hospital staffs. Morison did however show that phages, when taken orally, could prevent infection with cholera. This example would come close to modern, controlled, double-blind tests. Morison selected two regions in what is now Bangladesh to study the effectiveness of phages against cholera: Naogaon and Habiganj, which both had histories of serious epidemics. When the district official in Habiganj refused to participate in the trial, his region unwittingly became the control. Throughout the duration of the study, 1930 through 1935, there were no cholera epidemics in Naogaon, whereas Habiganj suffered severe outbreaks in 1930 and 1931, at which point the local government ordered the use of phages.[46] From then until 1935 the two districts had similarly low death rates from cholera, around ten per year, whereas in the nearby district of Sunanganj, where no phages were used, 1,505 people died from cholera in 1933 alone.[47]

Word Spreads

By this time, many scientists had taken up phage work. D'Herelle's discovery had intrigued researchers from Germany to Japan to South America, and they promptly began conducting experiments and reporting results of their own on illnesses like typhoid, skin infections and blood poisoning. One of the great enthusiasts of these early years was J. da Costa Cruz, an epidemiologist at the Institute Oswaldo Cruz in Rio de Janiero. By 1924, da Costa Cruz had prepared 10,000 ampoules of phage for distribution to hospitals and physicians across the country to fight dysentery, making it a routine treatment in Brazil. As a rule, he reported, patients began their recoveries 24–48 h after being treated. "The promptness with which the patient responds to the treatment is a matter of astonishment to the physicians," claimed one American

---

[44] C. A. Gill to L. Rogers, 18 August 1927, Item 58, Box C10, Leonard Rogers Correspondence, Wellcome Institute, London, cited in Summers, *Felix d'Herelle*, 133.

[45] Summers, *Felix d'Herelle*, 129.

[46] Morison, *Treatment and prevention of cholera*, 28; Morison, *Bacteriophage in cholera*, 563–570; d'Herelle, Perigrinations d'Un Microbiologist, 652, cited in Summers, 137.

[47] Ibid., 137.

doctor familiar with the campaign.[48] Yearly citations for papers on the therapeutic use of phage in the Index Medicus, the main international guide to medical literature, jumped tenfold between 1923 and the early 1930s, when experimentation with therapeutic phages apparently reached its peak. By 1932, wrote E. W. Schultz, some 1,200 papers had been published on the phenomenon.[49]

Laboratories like Schultz's, where scientists distributed phages to physicians in exchange for detailed case studies, opened at Columbia University, under the direction of bacteriologist Ward MacNeal, and at the Michigan Department of Health under Newton Larkum, who had devoted his Yale dissertation to the subject. In 1932, Larkum estimated that his lab had produced about 10 L of phage in the last 2 years, and that there existed about a score of so-called "private distributors" who, like him, furnished phage for experimental purposes.[50] In addition, three pharmaceutical companies, Eli Lilly, E. R. Squibb & Sons and Swan-Myers, a division of Abbott Laboratories, were producing various phage products – all unlicensed and never approved by the then-fledgling Food and Drug Administration. In fact, a study published in 1933 showed that most of these preparations were inactive.[51] The products came in the form of glass vials of liquid and tubes of jelly bearing names like "Staphylo-jel," "Strepto-jel" and "Colo-lysate."[52]

Many investigators who had tried phages on humans claimed excellent results. In 1929, Larkum wrote that the bacteriophage "possesses about all the virtues that could be desired of a therapeutic agent,"[53] because it readily destroyed bacteria, had no side effects and appeared to stimulate the body's own production of antibodies. In 1930, two physicians from San Antonio, Texas – E. D. Crutchfield and B. F. Stout – claimed a success rate of more than 90% in treating 57 patients with skin infections caused by staph bacteria.[54] That same year, Thurman Rice, an associate professor of bacteriology at the Indiana University School of Medicine reported a similar success rate in 300 patients suffering from various suppurative conditions – infections like abscesses, infected wounds and peritonitis where pus is formed.[55] He achieved

[48] E. W. Schultz, "The Bacteriophage: It's Prophylactic and Therapeutic Value," California and Western Medicine, Oct. 1927, vol. 27, no. 4. pg. 484.

[49] E. W. Schultz. 1932. "Bacteriophage As a Therapeutic Agent in Genito-Urinary Infections," California and Western Medicine, Vol. 36, No. 1 (Jan.), pg. 34.

[50] N. W. Larkum, "Bacteriophage in Clinical Medicine," The Journal of Laboratory and Clinical Medicine, vol. 17, no. 7, April 1932, p. 675.

[51] Margaret E. Straub and Martha Applebaum, "Studies on Commercial Bacteriophage Products," JAMA, Jan. 14, 1933, vol. 100, no. 2, pg. 110.

[52] Frederick Fitzherbert Boyce, Ralph Lampert and Elizabeth McFetridge, "Bacteriophagy in the Treatment of Superficial and Deep Tissues," New Orleans Med. And S. J. 86: 158–165, Sept. 1933.

[53] N. W. Larkum, "Bacteriophage Treatment of Taphylococcus Infections," Journal of Infectious Disease, vol. 45, p. 34, 1929.

[54] E. D. Crutchfield and B. F. Stout, "Treatment of Staphylococcic Infections of the skin by the bacteriophage, Archives of Dermatology and Syphilology, vol. TK, p. 1012–21.

[55] Thurman B. Rice, "The Use of Bacteriophage Filtrates in the Treatment of suppurative Conditions," American Journal of Med. Sci., vol. 179, pg. 346, 1930.

his best results in ten separate cases of children who came to the hospital with their bodies covered in boils – a possible result of malnutrition. "All of these have shown spectacular improvement immediately after the application of the bacteriophage," he wrote. "Most of these children were in bad condition when we began treatment – emaciated, badly nourished, running considerable temperature, extremely uncomfortable, and in several instances considered to be critically ill. The bacteriophage was applied directly as a wet dressing, or in several instances injected directly into the lesions with a fine needle. In every case except one the child was markedly improved the following day, the temperature was lower, the pain and soreness less, the early boils tending to abort, the older ones moving rapidly toward liquefaction, and the child apparently turned toward recovery."

Ward MacNeal of Columbia University claimed to have saved 7 out of 15 patients suffering from a severe bacterial blood infection by administering phages locally (in cases of infected wounds or boils), subcutaneously and by IV.[56] Before the advent of phage therapy, he wrote, patients in a similar condition invariably perished. Schultz reported some less dramatic but still encouraging results, based on reports sent back to him by physicians. In cases of chronic urinary tract infections, he reported a success rate of only 48%, while in acute cases, he found a far better rate of 87%. The inconsistent results, Schultz believed, invited skepticism. "Has bacteriophage a place among therapeutic agents of value to the urologist? I must admit at once that I have failed to bring you a categorical answer to this question," he said upon presenting these results to the Western Branch of the American Urological Association. "To some, this may represent the equivalent of a negative reply, but this is not necessarily the case. With any therapeutic procedure which does not yield uniformly successful results – and these are rare – one should not allow himself to be led astray by the failures which may initiate or sprinkle an inquiry." He concluded, "In closing, may I state that there is little doubt in my mind that in certain types of infections, bacteriophage, properly chosen and properly administered, is a therapeutic agent worthy of trial."

Phage therapy made headlines in the popular press as well. A story that appeared in the American journal *The Literary Digest* in December 1923 introduced the public to what it called "a mysterious product of nature's laboratory." The piece went on to summarize the results of one of the first phage studies conducted on patients in the United States. Two physicians at the Baylor University College of Medicine in Dallas had treated 20 children suffering from dysentery with phage and left another 12 untreated, as a control group. Of the 20 who were treated, 18 (90%), the authors claim, were cured, while of those left untreated, only 7 (60%) survived.

> The twenty children [who were treated with phage], varying in age from 4 months to six and half years, had been ill on the average for a little longer than 3 days with the deadly 'bacillary dysentery.' Their average maximum fever had been a little more than 103°. From the beginning of treatment to complete recovery was less than 6 days on the average, and there were no recurrences. In cases not treated by the bacteriophage, and which recovered, the average time required for recovery was 19 days.

---

[56] MacNeal, W. J. and Frisbee, Frances C.: Bacteriophage as a Therapeutic Agent in Staphylococcus Bacteremia, Journal of the American Medical Association 99: 1150–1155 (Oct. 1) 1932.

The study went so far as to provide a justification for the deaths of the two children who had been treated with phage: "Of these," claimed the authors, "one was a 'cretin,' an idiot dwarf as the result of an insufficient thyroid gland; the other was apparently on the way to recovery, when it ate green apples and died in convulsions."[57] The story ended with an encouraging quote from the physicians, Drs. Ralph C. Spence and Earl B. McKinley: "The bacteriophage holds enormous possibilities as a new weapon for fighting infectious disease."

The public was transfixed by the notion that there could be a living organism even smaller than a bacterium. In the lab, scientists used a ceramic filter, called a Chamberland filter, to sterilize solutions, and it was thought that no living organism would be tiny enough to pass through it. Though other so-called filterable viruses had been discovered by this time – ones that cause mosaic disease in tobacco plants, polio, measles and foot and mouth disease, among others – there was no consensus in the scientific community that viruses were in fact living things. "All that we knew, then, was that beneath the limits of a microscope's visibility, there existed 'something' and that this something was the cause of various diseases that affected plants, insects, birds, mammals," wrote d'Herelle.[58] "We knew the limits of a terra incognita into which no one had penetrated. I traveled through this land of the invisible, I discovered the most curious of its inhabitants, the parasite of microbes … I described its habits, its behavior, its comportment, without ever having seen it."

The idea that bacteriophages were alive ignited the public's imagination. "What is the smallest living being? The microbe? Microbes have lesser microbes that prey upon them," reported the New York Times in September 1925.[59] Many stories, like this one from Newsweek in 1933, quoted a then-popular ditty:

> Great fleas have little fleas
> Upon their backs to bite 'em,
> And little fleas have lesser fleas,
> And so ad infinitum.
> Last in the scale of little bugs comes the bacteriophage (bacteria eater). He is so tiny that he will slip through porcelain filters that trip up most bacteria, and he cannot be seen through the highest powered microscope.[60]

Art Imitates Life

No writer contributed as much to the popular fascination with phage therapy as Sinclair Lewis. His "Arrowsmith," which tells the story of a young doctor's journey through the parallel worlds of medical research and medical practice, must have read to some like science fiction when it was published in 1925, well before word of

---

[57] The Literary Digest, Dec. 22, 1923. Vol. 79, pg. 24.

[58] Felix d'Herelle, Perigrinations d'Un Microbiologist," 277, Archives de l"Institut Pasteur.

[59] "Tiny and Deadly Bacillus has Enemies Still Smaller," New York Times, Sept. 27, 1925, p. 14.

[60] "'Bugs Devour Deadly Disease Germs,'" Newsweek, June 10, 1993, p. 24.

d'Herelle's discovery had spread to the masses.[61] The book was a tremendous success, inspiring an entire generation of medical students before eventually falling entirely off school reading lists. D'Herelle himself read the book as soon as it was published and seemed delighted with it: "Have you seen the novel Arrowsmith by Sinclair Lewis? It is rather entertaining since it is almost entirely based on bacteriophage!"[62]

Lewis learned of d'Herelle through his friend Paul de Kruif, a scientist-turned-medical writer whom he had met at a dinner with mutual friends in 1922. They reportedly bonded over a mutual fondness for hard drinking.[63] De Kruif's writing informed much of Arrowsmith and provides valuable insights into the medical culture of the day. De Kruif, a bacteriologist by profession, had worked at the Pasteur Institute during World War I, though it's unclear if he ever met d'Herelle. He then served as a postdoctoral fellow at New York's Rockefeller Institute, a teaching hospital and the first center for clinical research in the United States. There, he quickly grew disillusioned with his colleagues and aired his complaints in his first published work, a series of anonymous articles in the journal *The Century*. The stories mourned the demise of the old-fashioned general practitioner, who carried out thorough physical exams without relying on a battery of blood work. He also criticized institutions like the Rockefeller for cultivating a "wild desire for priority" among its research staff. "The majority of workers are in a constant state of fear that some colleague will rush into print just ahead of them," he wrote. "This phobia results in an unnatural, strained effort and in a fatuous spirit of competition. Work is done under a pressure which usually results in researches that are incomplete, hurried, and botched."[64] It was a theme that would surface prominently in Arrowsmith. Shortly after the unflattering portrayal was published, de Kruif was forced to resign and threw himself into writing full-time.

Part muckraker, part service journalist, de Kruif celebrated heroic scientists, railed against medicine's shortcomings and strove to educate his readers about the importance of such things as routine checkups and good hygiene. One piece in Ladies' Home Journal celebrated the achievements of the nineteenth century Hungarian doctor Ignaz Semmelweis, who had discovered the link between poor hospital sanitation and childbed fever. De Kruif went on to report that 1 out of 18 mothers in America were still dying of the infection – a rate "higher than that of any civilized land." He urged expectant mothers to warn each other of this threat and to make sure their doctors scrubbed their hands before delivering their babies: "Women's oldest weapon – gossip – can end this worst of all scandals,"

---

[61] Summers, William C. "On the Origins of the Science in Arrowsmeth" pg. 316. July, 1991, The Journal of the History of Medicine and Allied Sciences, Inc. Vol. 46.

[62] Letter from d'Herelle to George H. Smith, dated 11 March 1925 from Alexandria, Egypt. Cited in Summers, Arrowsmith. Pg. 332.

[63] Summers, pg. 317.

[64] Paul de Kruif, "Our Medicine Men," pg. 953–954. The Century, vol. 104, 1922.

read the story's sub-headline.[65] Another piece informed readers of what he considered a promising treatment for pneumonia, a machine that sends a mild electric current through patients' lungs – and complained that it wasn't more widely available.

But the work that made de Kruif famous was his "Microbe Hunters," published 2 years after Arrowsmith, which turned infectious disease experts into international heroes. In vivid prose that approached the wide-eyed style of a children's storybook, he recounted the lives of such figures as Antonie Leeuwenhoek, the seventeenth century Dutch janitor who invented the microscope; Pasteur; Roux; and Paul Ehrlich, who in 1909 pioneered the use of salvarsan, the first chemically synthesized drug, against syphilis. Of Leeuwenhoek's discovery de Kruif wrote:

> This janitor of Delft had stolen upon and peeped into a fantastic sub-visible world of little things, creatures that had lived, had bred, had battled had died, completely hidden from and unknown to all men from the beginning of time. Beasts these were of a kind that ravaged and annihilated whole races of men ten million times larger than they were themselves. Beings these were, more terrible than fire-spitting dragons or hydra-headed monsters. They were silent assassins that murdered babes in warm cradles and kings in sheltered places.[66]

While working at the Pasteur Institute, de Kruif almost certainly learned of d'Herelle's discovery and may even have been present at the time of the breakthrough. Upon returning to the United States, he went to work at one of the first labs to conduct phage research, that of Frederick Novy at the University of Michigan.[67] Having been surrounded by superiors (Novy in the United States, Roux in France) who felt phages were significant, and having worked with five of the seven groups studying them, "de Kruif was almost uniquely positioned to bring this embryonic science into the popular mind," writes Summers in his study of the science behind Arrowsmith.[68] On the evening that de Kruif and Lewis met, the novelist was looking for a new subject for a book and de Kruif was still mastering the art of writing popular prose. The two decided to collaborate. In 1923 they traveled through the Caribbean – home of the fictional island where Lewis's hero, Dr. Martin Arrowsmith, would test his phages against plague – and plotted out the novel's framework.[69]

Aspects of Arrowsmith's life match exactly with d'Herelle's. Lewis wrote "D'Harelle [sic] 1917" in his book notes and considered naming his hero Felix.[70] Like d'Herelle, Arrowsmith discovers what he calls the "X Principal" just by chance. Rushing home, he tells his wife, Leora, "God, woman, I've got it! The real big stuff! I've found something … that eats bugs – dissolves 'em – kills 'em. May be a big

---

[65] Paul de Kruif, "Saver of Mothers," Ladies' Home Journal, vol. 49, March 1932, pgs. 6, 7, 124, 125.

[66] De Kruif, Paul. "Microbe Hunters," Harcourt, Inc. San Diego, New York, London. 1926. pg. 9.

[67] M. S. Marshall, "Observations of D'Herelle's bacteriophage," J. Infect. Dis., 1925, 37, 126–60, cited in Summers "On the Origins of the Science in Arrowsmith."

[68] Summers, Arrowsmith, pg. 319.

[69] Summers, William C. "On the Origins of the Science in Arrowsmith", 317.

[70] Summers, Arrowsmith, 318.

new step in therapeutics."[71] When word of his discovery spreads through the New York research institute where he's working, his superiors are overjoyed. "My dear boy," responds one of them, "I don't believe you quite realize you may have hit on the supreme way to kill pathogenic bacteria …!"[72] But in a twist on d'Herelle's own fate, before Arrowsmith has the chance to publish his first paper, another scientist beats him to it: d'Herelle. Arrowsmith's institute director breaks the news to him. "Something sort of bad – perhaps not altogether bad – has happened. … D'Herelle of the Pasteur Institute has just now published in the *Comptes Rendus, Academie des Sciences*, a report – it is your X Principle, absolute. Only he calls it 'bacteriophage.'" After Arrowsmith expresses bitter disappointment, his beloved director reminds him that science is about hard work and "not car[ing] – too much – if someone else gets the credit."[73] It's one of the novel's morality lessons for budding physicians and a direct echo of de Kruif's words.

Eventually Dr. Arrowsmith, like d'Herelle, gets the chance to try his phages in an exotic location: the fictional Caribbean island of St. Hubert, where citizens are dying of bubonic plague by the thousands. His mentor and boss, Max Gottlieb, gives him permission to go with the understanding that Arrowsmith observe rigorous testing standards.

> [Gottlieb] summoned Martin and remarked:
>
> If I could trust you, Martin, to use the phage with only half your patients and keep the others as controls, under normal hygienic conditions, but without the phages, then you could make an absolute determination of its value as complete as what we have of mosquito transmission of yellow fever, and then I would send you down to St. Hubert. What do you t'ink?
>
> Martin swore by Jacques Loeb that he would observe test conditions; he would determine forever the value of phage by the contrast between patients treated and untreated and so, perhaps, end all plague forever; he would harden his heart and keep clear his eyes.[74]

At the time, few doctors observed controls, because it was seen as unethical to deprive any patient of a promising treatment. De Kruif had railed against such "sentimentalism" in his 1922 book "Our Medicine Men."

> It has been pointed out that the application of the experimental method to human cases is little practiced in America, for two reasons. First because the training even of so-called scientific physicians does not demand the rigorous accuracy of carefully controlled experiments. Again, a popular hue and cry would certainly be raised against hospitals known to withhold a treatment from patients who might *possibly* be cured by it. As one great American physician remarked, it would not be justifiable to withhold from any patient a remedy that *might* be beneficial to save his life. But here, to speak colloquially, is the catch. It is, in the first place, necessary *accurately* to determine whether or not it is beneficial. For this the experimental method is demanded. This sentimentalism is short-sighted and foolish, and should be more vigorously combated by physicians, especially those who have pretensions to science.[75]

---

[71] Lewis, Sinclair, Arrowsmith, Penguin Putnam, Inc. New York, 1998, pg. 312.

[72] Lewis, Arrowsmith, 320.

[73] Lewis, Arrowsmith, 328.

[74] Lewis, Arrowsmith, pg. 348.

[75] De Kruif, Paul, Our Medicine Men, The Century Co. New York, 1922. pg. 96.

But the novel takes a more complex and humane approach to the issue. Many of Arrowsmith's colleagues object to his plans, including his friend Gustaf Sondelius, who refuses to be injected with phages unless Arrowsmith promises to similarly inject every island resident. Sondelius eventually dies, but Arrowsmith continues to push for controlled studies, even as islanders begin throwing rocks at him for withholding his treatment. Echoing de Kruif's writing, Arrowsmith boasts, "I am not a sentimentalist; I'm a scientist!"[76] But then, moved by the death of his own wife from plague, Arrowsmith defies his boss's strict orders and administers phage to every plague sufferer he encounters, thus failing to establish the unequivocal value of his discovery. It was an eerie echo of what was to become of phage in the real world.

Paris to New Haven and Back

Riding this crest of publicity, d'Herelle landed at Yale. "President Angell of Yale University today announced that Dr. *(sic.)* F. d'Herelle of Paris, distinguished in medical and bacteriological research and discoverer of bacteriophage, has been appointed Professor of Bacteriology in the Yale School of Medicine," reported the New York Times on June 15, 1928.[77] This would be the only permanent position d'Herelle would occupy in a top scientific institution. He was given lab space on the third floor of the Brady Memorial Laboratory, a modern, red brick building, and lived in various modest apartments near campus. His tenure, however, was marked by turbulence. He feuded with the dean of the School of Medicine, Milton Winternitz, who was unhappy with d'Herelle's frequent travels. D'Herelle had insisted on spending summers in France with his family and was a coveted lecturer at universities and medical societies across the United States, so he was frequently on the road. Winternitz also caught wind of the fact that d'Herelle was planning to launch a commercial lab in France and he strongly disapproved of the plan. Today, professors are typically encouraged to start their own companies, but back then the practice was frowned upon and seen as a distraction from more important academic commitments. D'Herelle, too, was ambivalent about mixing science with profit, noting that most commercial phage preparations were useless. "I now declare that I am, and always will remain, a stranger to all commercial enterprises," he had written in his 1926 monograph "The Bacteriophage and its Behavior." But this aversion was outweighed by d'Herelle's sense of responsibility for his own discovery. He was the "guardian of the method" of phage therapy and felt he should help insure the quality and proper use of the preparations.[78]

---

[76] Lewis, pg. 382.

[77] "D'Herelle to Join Yale Medical School," New York Times, June 15, 1928, p. 32.

[78] Summers, 174.

D'Herelle set up his lab, Le Laboratoire du Bacteriophage, in the late 1920s after having been approached by a Paris physician and a French pharmaceutical company. It marketed phage for a variety of ailments – sinusitis, wound infections and dysentery.

In the summer of 1929 Winternitz went so far as to have d'Herelle followed to see if his suspicions were justified. He asked a Yale colleague visiting Paris to drop in on d'Herelle's lab and report back. The visitor was rebuffed by d'Herelle and returned to the United States with little information. The lab remained open, run by d'Herelles son-in-law, into the 1970s.

D'Herelle and Winternitz had other quarrels. In one letter, Winternitz described an unhappy encounter with d'Herelle in 1930, in which the scientist "did not greet me but burst forth at once in broken English, telling me that he had not been treated fairly here." The issue appeared to be money, and Winternitz agreed to give d'Herelle a retroactive raise as well as additional funds for laboratory staff and expenses.

As the Great Depression deepened, it may have added to d'Herelle's discontent. On March 4, 1933, the day that Franklin Delano Roosevelt was sworn into office, d'Herelle's wife, Marie, noted in her diary: "All the banks are closed. We couldn't withdraw more than $25. There are fears of a 'rush.'"[79] Meanwhile, the d'Herelles' home country was enjoying a peak of prosperity. France's state treasury, which had been empty in 1926 reported a surplus of 19 billion Francs for 1929. Just 812 people were out of work, as compared to the millions in the United States, Great Britain and Germany. "Those last golden years of the 1920s promised to stretch into the next decade," wrote the journalist William Shirer, who was living in Paris at the time. "Life was good."

D'Herelle, who by all accounts had an imperious manner, must have felt uncomfortable in the more informal atmosphere of the United States, even beyond his sharp differences with Winternitz. "He was a well respected man, and he took himself very seriously," remembers one American scientist who met d'Herelle at Stanford in the 1930s. "But he seemed to be a little bit bloated with self importance, as some Frenchmen seemed to be."[80] And maybe the couple had grown tired of being away from home. Though Marie had only lived with her husband occasionally during his 5 years at Yale, she sorely missed her grandchildren, and filled up the diary that she and Felix kept jointly with poignant wishes to be speedily reunited with her "enfants adorés."[81] Whatever the reasons, the d'Herelles left Yale in the spring of 1933. In his May 10, 1933 letter of resignation, he cited "continued misunderstandings" and the "economic situation" as reasons for his departure.[82]

---

[79] Diary of Marie d'Herelle, March 4, 1933. Archives, Institut Pasteur.

[80] Author interview with Sidney Raffel. 2001.

[81] Ibid.

[82] Summers, 159.

It may not have been a great loss to the United States. By this time, the 60-year-old d'Herelle had all but exhausted his intellectual capital. His greatest research works, like the 1926 "Bacteriophage and Its Behavior," were behind him, and he felt satisfied with his contributions to phage therapy.

The d'Herelles left New Haven for France on the brand new, 30,000-ton cabin liner Champlain, which was hailed as the fastest and most luxurious of its class, able to cross the Atlantic in 7 days. By late May, Marie d'Herelle was happily back with her family in Paris.

Chapter 2
Inside Stalin's Empire

Shortly before the d'Herelles left New Haven in 1933, Felix received a well-timed invitation from a distant land. It came from his old friend and protégé Georgi Eliava, a dashing bacteriologist from the Soviet Republic of Georgia, who wanted to enlist d'Herelle's help in establishing a bacteriophage institute in its capital, Tbilisi.

In the 1930s, the great centers of medical and scientific research were based in Germany, France, Great Britain and the United States; the Soviet Union was desperate to catch up. "We are 50 or 100 years behind the advanced countries," Joseph Stalin proclaimed in 1931. "We must make good this distance in 10 years. Either we do it, or they crush us."[1] To accomplish that goal, the Kremlin was setting the USSR on a course of rapid industrialization and luring key scientists and engineers from overseas. D'Herelle was one of them.

Pleased with the prospect of going where his work might be more fully appreciated and always looking for a new adventure – even at the age of 61 – d'Herelle readily accepted. D'Herelle had met Eliava, who was nearly 20 years d'Herelle's junior, during the older scientist's last and most stormy months at the Pasteur Institute. In what would become something of a pattern for d'Herelle, he had alienated a coworker, Albert Calmette, who would later become his boss. D'Herelle publicly expressed doubts about an anti-tuberculosis vaccine, called BCG, that Calmette had developed with a colleague. "Upon our first interview after my return from Indochina, [Calmette] asked if it was true that I considered BCG as dangerous," wrote d'Herelle in his memoirs referring to late 1920 or early 1921. "I replied that was correct … [B]eginning at that moment, he swore toward me an implacable hatred."[2] After being named deputy director of the Pasteur Institute, Calmette took his revenge by giving away d'Herelle's laboratory space to another scientist.[3] Fortunately a colleague, Edouard Pozerski, came to d'Herelle's aid by giving him

---

[1] Stalin: Great Lives Observed (Prentice-Hall, NJ) 1966, editor: T. H. Rigby, pg. 47.

[2] Felix d'Herelle, Perigrinations, 485, cited in Summers.

[3] Felix d'Herelle, Perigrinations, 485, cited in Summers.

A. Kuchment, *The Forgotten Cure: The Past and Future of Phage Therapy*,
DOI 10.1007/978-1-4614-0251-0_2, © Springer Science+Business Media, LLC 2012

the use of "a corner of a table in his minuscule laboratory." "It is there," d'Herelle wrote, "that I made the acquaintance of a young Georgian, Georges Eliava, who at first was my student and later remained my friend."[4]

The two had much in common. Eliava had arrived at the Pasteur Institute already captivated by phages; according to relatives and coworkers, Eliava had independently witnessed their action a few years before. "Once he left a [cholera] culture under a microscope on a slide," says Eliava's granddaughter, Natasha Maliyeva. "But, because he was easily distracted – often carried away with different ideas – he left it there and didn't open his lab again for an entire week. When he returned to the lab, the slide was clear. And he knew very well that he hadn't washed or cleaned the slide. When he repeated the experiment, the phenomenon repeated itself." Because Eliava never published any papers on the phenomenon, Natasha's tale is difficult to prove – but is mentioned separately in an Eliava Institute document.[5]

Eliava also helped defend d'Herelle's discovery from attacks by coworkers. While d'Herelle was traveling in rural France, as he frequently did in late 1919 and early 1920, carrying out phage therapy field trials in birds afflicted with typhoid,[6] the debate over whether phages were living organisms erupted, with several prominent scientists coming out against d'Herelle's point of view. Meanwhile, Eliava asked permission from Institute Director Emile Roux to carry out some phage experiments of his own. According to Eliava's granddaughter, those experiments showed that d'Herelle was right. "So, d'Herelle was summoned from the French countryside," she says. "He was already despairing of being able to prove he was right. My mother told me that when he arrived and entered the institute he yelled out at the entrance, 'Where is this Eliava? Show him to me!' and when Gogi [Eliava's nickname] ran out to meet him, [d'Herelle] hugged him and they kissed. And from that point on they were like father and son."[7]

Whether or not the story is true – it isn't mentioned either in d'Herelle's memoirs or those of Pozerski – there's little doubt that the two shared a warm friendship and a productive collaboration. In 1921, they published two papers together, including one that directly challenged some of the claims made during d'Herelle's travels and argued that phages were living organisms. About that same time, d'Herelle terminated his relationship with the Pasteur Institute and Eliava returned to Georgia. "They need me there," he said.[8]

Eliava was as charming and laid back as d'Herelle was prickly and businesslike. A handsome, athletic figure with dark pomaded hair and small features, Eliava was described by Pasteur Institute coworkers as '*poli*,' a term that could have meant

---

[4] Ibid.

[5] Georgadze, pg. 18 cites a paper called "Viruses Against Microbes," by A. S. Kriviskii, MEDGIZ, 1962, which states, "In 1917, a young talented Georgian microbiologist G.G. Eliava, analyzing water from the Kura river for the presence of cholera, also confirmed this [d'Herelle's] result."

[6] Summers, 63.

[7] Author interview with Natasha Maliyeva, Tbilisi. Nov. 3, 2002e.

[8] Maliyeva.

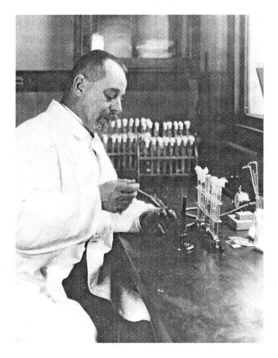

either polished, polite or smooth with women – all of which were true. He was married to one of the most glamorous figures in Tbilisi, Amelia Vol Levitskaya, a Polish-born soprano and a soloist with the Tbilisi State Opera, whom he had stolen away from an unhappy marriage in Warsaw. He had also adopted her young daughter, Hannah, whom he brought up as his own child. Pozerski recalled that Eliava and his family lead a somewhat lavish existence, having managed to stash money with friends in Paris before Communism engulfed their homeland. "Like a true Georgian," he wrote, "he loved horses and bought a purebred mare from a friend of mine who was a trainer, so he could bring her back to Georgia."[9]

Like d'Herelle, Eliava came from a privileged background. His roots were in the village of Sachkhere in western Georgia, where he was born in 1892, but he grew up in Batumi, the large port city on the Black Sea. His father was a prominent doctor from a well-connected family that counted the writer Leo Tolstoy among its friends. But while d'Herelle's formal education ended when he was a teenager, Eliava went on to complete medical school. He began his studies at Odessa University in 1909, where he majored in literature, but he was soon expelled for participating in an underground student movement and exiled back to Georgia under police supervision. In 1912, he moved to Switzerland to attend Geneva University, where a series of lectures on bacteriology captured his interest. While he was home on summer break in 1914, World War I broke out and prevented him from returning abroad.

---

[9] Edouard Pozerski, *Souvenirs d'un demi-siecle a l'Institut Pasteur.* Pasteur Museum, 46.

With the help of an influential relative, he was accepted to Moscow University, where he studied medicine and finally graduated in 1916 – just 1 year before the Russian Revolution unseated the czars and lead to Georgia's shortlived independence as the Georgian Democratic Republic (1918–1921).[10]

After completing his studies, Eliava was mobilized and stationed in Trabzond, in modern-day Turkey, where he headed a microbiological lab that worked to prevent and contain disease outbreaks within the Russian Imperial Army as it battled troops from the Ottoman Empire. By 1918, Eliava had moved to Tbilisi, then called by its old name, Tiflis, to take over the Tiflis Central Laboratory.[11] It was that lab's governing body, the League of Cities, that would send Eliava to Paris to learn vaccine and serum production and to purchase medical equipment for the newly independent Georgian government.[12, 13]

The Pasteur Institute had by then become a Mecca for medical professionals from around the world, and scientists from the Russian empire were no exception. As early as 1888, just 3 years after Pasteur completed the first human trials for his rabies vaccine, a Tbilisi physician went to Paris to learn about the life-saving medicine. He brought back two rabies-infected rabbits for use in vaccine production, and opened an anti-rabies lab in the center of Tbilisi that same year.[14] The lab, called a "Pasteur station" in Russian, would eventually be subsumed into the institute that Eliava and d'Herelle would found together.

Eliava arrived in Paris against the backdrop of terrifying epidemics and famine in his homeland. Even before 1914, health care in Russia lagged far behind that available in the West. The outbreak of World War I only made matters worse.[15] Typhus, a deadly fever transmitted by body lice, smallpox and the worldwide pandemic of influenza were just three of the killers that preyed upon an already weakened, exhausted population. Between 1916 and 1924, some ten million people succumbed to infectious diseases. When the Bolsheviks came to power in 1917,

---

[10] Georgia Medical Journal, March–April 1992.

[11] Article: "The Tbilisi Scientific Research Institute of Vaccines and Serums, Ministry of Public Health, USSR, Over the Last 50 Years," by N.A. Georgadze, From the book: "Theoretical and Practical Questions of Bacteriophagy," Eds, N.A. Georgadze, et al. Tbilisi, The Tbilisi Scientific-Research Institute of Vaccines and Serums, 1974, pgs. 9 and 10.

[12] Letter signed by Assistant Minister T. Karsivadze, Georgian National Historic Archive, file #625, roll 92, box 28, cited in "Returned Names," Medical Journal of Georgia, March–April 1992.

[13] Certificate signed by N. Eliava of the Central Committee of the League of Cities of the Republic of Georgia, Sept. 31, 1919, cited in "Returned Names," Medical Journal of Georgia, March–April 1992.

[14] Article: "The Tbilisi Scientific Research Institute of Vaccines and Serums, Ministry of Public Health, USSR, Over the Last 50 Years," by N.A. Georgadze, From the book: "Theoretical and Practical Questions of Bacteriophagy," Eds, N.A. Georgadze, et al. Tbilisi, The Tbilisi Scientific-Research Institute of Vaccines and Serums, 1974, pg. 10.

[15] Weissman, Neil B., "Origins of Soviet Health Administration," in the book Health and Society in Revolutionary Russia, Susan Gross Solomon and John F. Hutchinson, eds., Indiana University Press, Bloomington and Indianapolis, pg. 102.

Lenin made public health a top priority. "This typhus," said Lenin shortly after the Revolution, "in a population [already] weakened by hunger and sickness, without bread, soap, fuel, may become such a scourge as not to give us an opportunity to undertake socialist construction. This [must] be the first step in our struggle for culture and for [our] existence." At the height of the epidemic in 1919, he declared before the Seventh Congress of Soviets: "Either the lice defeat socialism or socialism defeats the lice." One year later, he swore to turn all of Russia's resources toward the fight against epidemics.[16]

Eliava arrived back in Tbilisi in November 1921 with boxes of lab equipment, vaccines and serums worth nearly 100,000 francs. He soon managed to secure enough money from the ministry of public health to expand his lab into the Tiflis Bacteriological Institute, where he introduced a research program on bacteriophage and began the manufacture of vaccines for cholera, smallpox and typhus. For the first time, Georgia was able to manufacture its own medicines instead of having to import them.[17] After another short stint at the Pasteur Institute, from 1925 to 1927, Eliava became chairman of the department of microbiology at Tbilisi University.[18] But he never forgot his old mentor or the thought that d'Herelle's bacteriophages might one day save thousands of lives in his homeland. Now, able to attract significant government resources for his projects, Eliava offered d'Herelle an appointment as professor at the school for continuing education of physicians and a consulting position with what he hoped would be a world center for the study of bacteriophages.

The d'Herelles arrived in Georgia (then a piece of the Transcaucasian Soviet Federated Socialist Republic) in October 1933 on a ship that had taken them from Marseilles across the Mediterranean, the Dardanelles and the Bosporus into the Black Sea. They docked in Batumi, Eliava's boyhood home, where their old friend met them with an armful of flowers.[19] At the time, Joseph Stalin had been in power since Lenin's death in 1924 and his cult of personality was already alive and well, with statues and portraits of the 5″ 4′ autocrat going up in offices and major public places across the country. Patriotic fervor was at a peak, with the Georgian-born leader (his birth name was Dzhugashvili) well on his way to turning the once backward land into an industrial and military megalith to rival the Western powers. New cities were built, mines were sunk, peasants were trained by the millions – and in many cases forced – to join the booming workforce, and engineering and educational institutions expanded. With the United States suffering under the Great Depression, Stalin was making a convincing case that the Russians had "overthrown capitalism".[20]

[16] Field, Mark G., "Soviet Socialized Medicine: An Introduction." The Free Press, New York, pgs. 51–52.

[17] Georgadze pg. 13; Summers, pg. 162.

[18] Who Was Who in the USSR, eds. Dr. Hienrich E. Schulz, Paul K. Urban, Andrew I. Lebed, the Scarecrow Press, Inc. Metuchen, NJ 1972, 155.

[19] D. P. Shrayer, "Felix D'Herelle in Russia," *Bull. Inst. Pasteur*, 1996, vol. 94, pg. 92 and 93.

[20] Service, Robert. Stalin: A Biography. The Belknap Press of Harvard University Press, Cambridge, Mass. 2005. 265–275.

The d'Herelles would stay in Georgia until April 1934 and return for another short stay from November 1934 until the following May. Though they would soon experience the dark side of Stalin's rule, the d'Herelles at first noted mostly the positive changes that Communism brought to cities across the Soviet Union. Tbilisi's twisty, medieval streets were widened and straightened to ease traffic flow, new parks were built, trees lined the boulevards and the rural outskirts were linked to the center with new roads. It was all part of an effort to beautify urban areas and make them habitable for the working classes the country glorified. "The new square is resplendent and surrounded by stores," wrote Marie upon returning to Tbilisi in November 1934. "There are perfumeries, pretty pharmacies, cafes, a large ready-to-wear store for men. All the buildings have had a facelift, which gives the city a nicer [appearance?] than last year."[21] On another day she noted: "At noon we took a walk that lasted until 2 o'clock. So many changes. They have built beautiful quays along the Kura and a pretty promenade bordered by trees."[22]

During d'Herelle's visit, he and Eliava worked hard to lay the foundations for the future bacteriophage institute. Their coworkers remember how d'Herelle would arrive at the institute no later than 8 a.m. each day. "For us young lab workers ... it was hard to keep up with him," recalled Irakli Georgadze, who would became director of the institute in 1959.[23] Georgadze recalled that d'Herelle was a "virtuoso" in the lab and insisted on doing everything with his own hands – accepting little help from assistants and technicians. One day, after a lab worker forgot to sterilize a batch of glass pipettes, d'Herelle hand blew his own over a flame. "There was no end to our wonderment," wrote Georgadze.[24]

D'Herelle also brought with him from France and donated, much of it at his own expense, laboratory equipment for the new center. Even after he returned to Paris in 1935, he continued to send much-needed equipment to Tbilisi, including a handsome wooden desk with a white ceramic top and a wooden cabinet that houses an incubator for growing cultures at 37°C.[25] D'Herelle traveled the Soviet Union, occasionally giving lectures on bacteriophagy and meeting with physicians interested in the treatment. Sometime between December 1934 and February 1935,[26] d'Herelle also met with Grigoriy Kaminsky, the Commissioner of Public Health for the Soviet Union. Kaminsky invited d'Herelle to continue his research in Moscow, where he offered to devote one scientific institute entirely to the study of phage therapy.[27] D'Herelle declined for a variety of reasons, including health: the 61 year old scientist

[21] Ibid. Nov. 8, 1934.

[22] Ibid. Nov. 28, 1934.

[23] Georgadze, pg. 22.

[24] Georgadze, pg. 23.

[25] I. Georgadze, personal communication to David Shrayer, 1977, cited in Shrayer, Felix D'Herelle in Russia, 95.

[26] The d'Herelles' diary gives the date of December, 1934. Georgadze dates the meeting to January–February 1935.

[27] Georgadze, I. A. (1974) Fifty years' sum of Tbilisi Research Institute of Vaccines and Serums, in Theoretical and Practical Aspects of Bacteriophage (pg. 21). Tbilisi, (Russ.).

suffered from unspecified respiratory problems that he feared would be aggravated by Moscow's inhospitable climate.

While in Tbilisi, D'Herelle also completed a third summary of his work on phages, a book that was translated into Russian and Georgian by Eliava and published in 1935. It was dedicated to Stalin. The introduction, seemingly penned by d'Herelle, heartily praises Soviet science and calls the USSR "a remarkable country, which, for the first time in the history of mankind, chose as its guide not irrational mysticism, but a sober science without which there cannot be any logic or genuine progress." The words appear to be "d'Herelle's own scientific manifesto upon starting his life and work in the Soviet Union," writes d'Herelle's biographer, Summers.[28]

The trip wasn't all work. The d'Herelles spent much of their leisure time with the Eliavas, going on excursions or taking in Mrs. Eliava's solo performances with the Tbilisi State Opera. Marie thought her voice *"superbe."*[29] They must have mingled with some of the Eliavas' illustrious circle of friends that included many of the top artists, writers and performers of the day. Felix d'Herelle, a talented amateur photographer, brought back rolls of black and white film from their trip. The photos, now in the archives of the Pasteur Institute, show the d'Herelles and the Eliavas, wearing elegant hats, the women invariably in fur coats, sightseeing in Tbilisi, Batumi, Baku, Leningrad and other parts of the USSR.

But the photos must have masked a growing level of unease. The d'Herelles had arrived during a time of mounting repression across the Soviet Union that would culminate in the Great Purge of the mid-to-late 1930s. Though Stalin's economic policies were crucial to the country's future victory in World War II, they had inflicted misery on large pockets of the population. Millions of peasants had died during forced collectivization (those who refused to participate were shot or sent to

---

[28] Summers, pg. 165.

[29] Diary of Marie and Felix d'Herelle, Nov. 13.

labor camps) and the overall standard of living had plummeted. The leader began to fear a civil war – especially, one aided from abroad. "We have internal enemies. We have external enemies. This, comrades, must not be forgotten for a single moment," proclaimed Stalin. Beginning in 1928, a series of defendants were publicly tried and convicted in Moscow for alleged conspiracies against the State. Among the accused were a group of economic officials who had allegedly undermined the Soviet economy on instructions from France and some 53 engineers from the town of Shakhty accused of espionage and "wrecking," a term that meant something akin to sabotage. Each trial was reported in the press as a great public event, and the accusations were invariably described as true. Over the next several years scores of 'bourgeois specialists' were removed in every field: teachers, technicians and anyone who had been trained or educated under the pre-Revolutionary regime. The motives for the killings were manifold: to maximize Stalin's personal power and security, to provide scapegoats for economic failure, and to increase xenophobia.[30, 31]

Little did the d'Herelles know that their close friend, Eliava, was in danger. In 1933 – presumably before the d'Herelles' arrival that October – Eliava had been detained with some 17 other prominent Georgians on charges of wrecking, sabotage, espionage, and of belonging to a so-called Georgian National Center, which, the authorities claimed, was in charge of all counterrevolutionary activity in the territory.[32] Eliava, unfortunately, fit the typical profile of a target of persecution: not only was he educated before the Revolution but he had studied abroad, came from an aristocratic family, was an intellectual and maintained ties with foreigners. According to the memoirs of a former member of the Transcaucasian secret police, Suren Gazaryan, the group was arrested on direct orders from the infamous Lavrenti Beria, then the first secretary of the Transcaucasian communist party and future chief of the Soviet secret police. They were kept in a single holding area and forced to confess to absurd activities. "Some of those confessions, by their fantastic nature, resembled fairy tales from A Thousand and One Nights," writes Gazaryan. Eventually, the group was freed after Gazaryan and his coworkers brought the dubious case to the attention of authorities in Moscow.

But Eliava's troubles did not end there. In July 1936 civil war broke out in Spain, pushing Stalin to new heights of paranoia. "He was shocked by the ease with which it had been possible for General Franco to pick up followers ..." writes Stalin biographer Robert Service. "He intended to prevent this from ever happening in the USSR."[33] The following month several Soviet leaders, including Beria, were ordered to pen articles that would set the stage for the Terror. Beria's piece, which appeared in the newspaper *Pravda*, was entitled "Scatter the Ashes of the Enemies of Socialism." In it, he urged every party organization to weed out "degenerates,"

---

[30] Conquest, 152–155.

[31] Service, 344.

[32] Suryan Gazaryan, 1989, *Eto ne dolzhno povtoritsa [This should not happen again]*. Zvezda 1, 3 (1989), Leningrad (Russ.).

[33] Service, 347–348.

"careerists," "self-seekers," "those without morals," or anyone else who might allow the "counterrevolutionary-espionage-Trotskyite-Zinovievite" to prosper. (Like Stalin, Beria was paranoid about foreign influence, having been profoundly affected by a 1924 Menshevik revolt in which Britain and France supported an uprising in Georgia against the Soviet Union.)[34] That same month, the first defendents – Grigory Zinoviev and Lev Kaminev, who had helped bring Stalin to power – were paraded in court, forced to confess to trumped up accusations, sentenced to death and shot. Two other major trials followed, allowing Stalin to consolidate power and tame the Soviet Communist Party through terror. The trials were the public face of a secret, semi-haphazard, out-of-control persecution that spread throughout the country and killed millions of innocent victims.

On a more local level, Eliava's coworkers testified to a personal rivalry between him and Beria. Some say Beria felt snubbed by Eliava. When Eliava completed his translation of d'Herelle's book, he sent it directly to Stalin through a friend of his, the prominent Moscow party leader Sergo Ordzhonikidze. In doing this, he went over the head of Beria, the political boss of Transcaucasia.[35] Eliava may have also gone over Beria's head in seeking permission to establish the Bacteriophage Institute. His granddaughter says he applied to Moscow with his request through another friend of his, Budu Mdivani, the first deputy chairman of the Georgian Council of People's Commissars who would be executed at the height of the purges

---

[34] Razveiat' vprak vragov sotsializma," Pravda, 19 August 1936, cited in Knight, Amy. "Beria: Stalin's First Lieutenant." (Princeton: Princeton University Press, 1993), pg. 72.

[35] Shrayer, 94.

in Georgia in 1937. "Eliava, through [Mdivani], sent a letter to Stalin where he very seriously stated why and for what purpose he wanted to establish this institute in Tbilisi, what was his perspective," Maliyeva said. "And he got permission over the head of Beria. From Moscow, Stalin personally ordered Beria to sign papers and make all the necessary provisions for Eliava. That, of course, was unforgivable."

They may have also clashed over women. The balding, bespectacled Beria was known for his sexual debauchery – when he later moved to Moscow, Stalin wouldn't trust Beria around Stalin's own teenage daughter.[36] Eliava, handsome and flirtatious, might easily have stepped on Beria's toes.[37] It's more likely, however, that Beria was merely filling quotas for Moscow. A Stalin lackey known for his cruelty – Beria kept a cane in his office with which to beat victims during interrogations – he would stop at nothing to curry favor with his boss.

But whatever the reasons for the enmity between the two, Beria would soon gain unprecedented power. On Dec. 5, 1936, with the adoption of Stalin's new "Great Constitution of the USSR," the Transcaucasian Soviet Federated Socialist Republic was broken up into the individual republics of Georgia, Armenia and Azerbaijan, and Beria became Georgia's supreme leader. Just 1 month later, all 17 people who had been freed in 1933 were rearrested and ordered shot.

On January 22,[38] 1937, Eliava and "Mira," as he affectionately called Amelia, had just returned home, exhausted, from a day-long outing that had included a visit to Saburtalo, where the stables were nearly complete. This was where Eliava kept his prized horse, Infanta,[39] and where scientists would later keep the horses they would use for serum production. The Eliavas had a dinner guest – a new friend who had recently started coming over regularly and who Hannah Eliava realized in retrospect was probably working for the KGB. Just then, a group of soldiers came in and declared that they were placing Eliava under arrest. "Take care of your mother," he told Hannah before being led away, never thinking that his wife, too, would be arrested moments later. Hannah, who was 23, remained alone in the apartment with her fiancé while the soldiers conducted a thorough search of the premises. In April, Hannah, too, was put in jail. There, she saw the flower of Georgia's intelligentsia wasting away in inhuman conditions – crowded 50 people to a single cell.

One day, the door opened and a new prisoner was brought in. To her horror, Hannah recognized her mother. The once beautiful, raven-haired singer had turned into an old woman. Emaciated, with swollen sacs under her eyes, she lifted her skirt to show Hannah how she had been beaten. Relatives were not allowed to share a cell, so the women's fellow prisoners protected them, taking turns looking out the peep hole to make sure no one was coming while mother and daughter stayed up all

---

[36] Service, 431. "When she [Svetlana] wanted to stay overnight at the Berias' dacha, where she was a frequent visitor, he ordered her to return home immediately: "I don't trust Beria!" Stalin was aware of Lavrenti Beria's proclivities toward young women."

[37] E. Tsereteli, personal communication to David Shrayer, cited in Shrayer, 94.

[38] This is the date firmly recalled by Eliava's family. Official documents give it as Jan. 20, 1937.

[39] Infanta is a Spanish word that in Russian is used to mean a princess who is heir to the throne.

night talking and crying. Ultimately, Amelia Vol-Levitskaya and Georgyi Eliava were shot, although the exact date of their execution is not known. Hanna was eventually transferred to a labor camp in Kazakhstan where she languished for 9 years. On July 11, 1937 the following report appeared in Georgia's main newspaper, Kommunisti:

"The scientist Georgyi Eliava has answered Stalin's fatherly devotion to scientific workers by conducting terrible wrecking.[40] On the orders of bloody fascist Trotskyites he was preparing bacteria to kill the Soviet people, especially in the case of war. This animal got the response of the Soviet people that he deserved."[41]

Word slowly trickled back to Paris about the horrors taking place in Tbilisi. By one account, d'Herelle was informed of his friend's arrest and urged the French government to intervene. A call was placed to Stalin, who called Beria, who replied that Eliava had already been shot. In fact, he was still alive, but Beria ordered him shot that same day.[42]

Others received different but equally unsubstantiated versions of the events. Pozerski, d'Herelle's former Pasteur Institute colleague, recorded the following in his memoirs, a story he heard from Georgian refugees:

One morning, a group of soldiers burst into the institute. They were under the command of an officer who, as soon as he entered the courtyard, spotted from afar the French thoroughbred, in a stable with the door open. He entered.

As a true Caucasian, he couldn't restrain himself from jumping on the mare. Reacting to the prod of a spur she had never felt, the horse reared up and, brusquely, bounded out of the stable with such speed that the officer didn't have time to duck. His forehead struck the frame of the open door. He fell backward, covered in blood, motionless, his forehead smashed.

All the soldiers dashed toward Eliava, took him away. His wife cried out: they've arrested him, they've taken him away.

What became of her? I never found out. But what I learned the following year, in Paris, from some Georgian refugees, is that Eliava was tried and shot that same day … and the mare too, for having caused, both of them, the death of a commandant.

Poor Eliava! His photograph still hangs on the wall of my laboratory, of his Paris laboratory, at the Institut Pasteur.

What a drama provoked by d'Herelle's bacteriophage, discovered at Paris's Institut Pasteur![43]

Until the time of Eliava's arrest, it seemed, d'Herelle had every intention of returning to the USSR to continue the joint project. "I remember very well seeing [d'Herelle] off for France with his wife Marie," recalls Irakli Georgadze, who would become the Institute's director. "… I remember Batumi port where the d'Herelles boarded the Italian ship 'Delmatzio.' A snow-white ship. A snow-white uniform on

---

[40] "wrecking" is the formal translation for a term meaning anti-party activity.

[41] Kommunisti No. 157 (4958), cited in forthcoming book of Avtandil Kurkhuli.

[42] Author interview with Revaz Adamia, Georgian Ambassador to the United Nations, August 22, 2003.

[43] Edouard Pozerski, *Souvenirs d'un demi-siecle a l'Institut Pasteur.* Pasteur Museum, 47. Eliava's granddaughter says the story is probably fictitious.

the captain of the ship. 'I am going to come back in the autumn, said professor d'Herelle. And you, Irakli, are going to be speaking French fluently by then, OK? Because I want you to go and work at the Pasteur Institute for a year or two. A real microbiologist should go through this school.'"[44]

The ship sailed off, and with it ended all ties between Tbilisi's Bacteriophage Institute and the West for more than 50 years. In the fall of 1935 d'Herelle was injured in a train accident and spent the next several months, until the spring of 1936, recuperating in the French Riviera.[45] By then, Stalin's Great Terror was underway and d'Herelle never returned.

The tragic end of d'Herelle and Eliava's partnership also marked the dimming of the prospects for phage therapy in the West. D'Herelle, by then in his 60s and living a fairly comfortable life in Paris, turned away from his tireless advocacy for phage therapy and toward summing up his life's accomplishments. During World War II, which he spent with his wife in Vichy, France – the capital of Phillipe Petain's Free Zone – he wrote two voluminous manuscripts that were never published and still sit in their original form at the archives of the Pasteur Institute in Paris: "Les Perigrinations [sic] d'Un Microbiologiste," his autobiography, and "La Valeur de l'Experience," his monograph describing his scientific ideas.[46] The Nobel Prize eluded him, but he did win some belated recognition from the scientific establishment just before his death. Over the protests of some of its older staff, in 1947 the Pasteur Institute commemorated the 30th anniversary of the publication of d'Herelle's first phage paper by inviting him to speak and presenting him with the Medal of the Pasteur Institute. The next year, the French Academy of Sciences awarded him the Prix Petit-d'Ormoy for natural science. One year later, he died of pancreatic cancer. The salmon-colored cottage where d'Herelle and Eliava were supposed to live became a part of the Georgian office of the KGB and remains off-limits to the Eliava Institute.

---

[44] I. Georgadze, personal communication, 1977, cited in Shrayer, 95.

[45] Summers, 165.

[46] Summers, 178–180.

Chapter 3
The Fading of Phage Therapy

While physicians in the United States continued to experiment with phage therapy through the 1930s, their successes were marred by flawed research. Patient trials were conducted without controls and, in many cases, on self-limiting conditions – infections like boils, acne and abscesses that typically heal on their own. Often, doctors opened the abscesses and drained wounds before flushing them out with bacteriophage, making it unclear if it was the surgical procedure, the phage or a combination of the two that had cured the infection. So, toward the mid-1930s came an increasingly urgent call for strictly controlled human trials. Wrote Newton Larkum of the Michigan Department of Health in 1932, "Because of conflicting experimental observations, enthusiastic and poorly controlled clinical application and rapidly expanding commercial exploitation, a situation is developing which will, unless guided and checked, lead to the ultimate rejection of bacteriophage by all who make any pretense to the practice of scientific medicine."[1]

It's unclear why this call for strict controls was never heeded. Possibly, it resulted from a lack of government intervention – the United States Food and Drug Administration would not start formally evaluating medicines for safety and efficacy on a wide scale until the late 1930s, with the passage of the Food, Drug and Cosmetic Act of 1938. At the time, many physicians may have felt uncomfortable with the idea of withholding a seemingly promising treatment from a group of patients, as required by double blind studies – the dilemma confronted by Lewis's Arrowsmith.

Nervous about the unchecked proliferation of commercial bacteriophage products and the growing number of physicians and patients using them, the American Medical Association weighed in with two influential reviews of the treatment, first in 1934 and again in 1941. The association had sounded a cautionary note on phages as early as 1929, when a critical editorial appeared in its journal. "The theory of

---

[1] N. W. Larckum, "Bacteriophage in Clinical Medicine," The Journal of Laboratory and Clinical Medicine, vol. 17, no. 7, April, 1932, pg. 675.

A. Kuchment, *The Forgotten Cure: The Past and Future of Phage Therapy*,
DOI 10.1007/978-1-4614-0251-0_3, © Springer Science+Business Media, LLC 2012

such a lytic principal has been sufficiently engrossing to enlist the interest of the writers of fiction," it said, referring to Arrowsmith. "But fact and fancy may be more conflicting in everyday life than in the world of imaginative literature."

The 1934 review showed just how conflicting the literature on phage therapy was. Scientists couldn't agree on how a bacteriophage worked inside a person's body, or if it worked at all. Many suggested that patients recovered not because phages destroyed the bacteria but because the presence of bacterial proteins in the phage serum set off a natural immune response, much like a vaccine. Others were puzzled by the seeming failure of investigators to demonstrate the efficacy of phage therapy in lab animals. "It has been the general experience in medical research that therapeutic agents which are effective in human disease have previously been found effective in suitably conducted laboratory experiments on animals. This is not the case in phage therapy," wrote the Journal of the American Medical Association study's authors, Monroe Eaton and Stanhope Bayne-Jones of Yale's department of bacteriology.

Many prominent researchers also continued to misunderstand what a bacteriophage was. The JAMA reviewers reached the incorrect conclusion that the bacteriophage is "inanimate, possibly an enzyme." While a few rising scientific stars, including Herman Muller, had understood that phages were viruses as early as 1922, they were in the minority. In a 1993 paper, Summers points out that, of five textbooks published between 1929 and 1936, only two argued for d'Herelle's (correct) notion that phages were parasites. The other three sided with d'Herelle's Pasteur Institute rival Bordet, who (incorrectly) claimed they were enzymes. "Even the two texts that favored the d'Herelle model presented it as a minority view, but likely true nonetheless," he writes. D'Herelle's inability to attract more scientists to his point of view was more a reflection of his own personal style than of the merits of his experimental abilities. As Summers notes, Bordet and other proponents of his views were prominent academics in important posts (besides Bordet, they included John Northrop of the Rockefeller Institute and Philip Hadley of the University of Michigan). And while these experts based their writing on complex principles of biochemistry and immunology, the self-taught d'Herelle tended to articulate his arguments in simpler terms, relying on a combination logic and name-dropping. In his book "The Bacteriophage and Its Behavior," he recalled that Albert Einstein himself had once expressed support for his view that a phage was a living organism. "I do not believe that there are a great many biological experiments whose nature satisfies a mathematician," he wrote.[2]

The JAMA review concluded that the literature on the use of phage is "for the most part contradictory," that the only diseases for which it could be of value were local skin infections and that the favorable results reported seemed to have been due not to the bacteriophage but to "the specific immunizing action of bacterial proteins in the

---

[2] Summers, William. "How Bacteriophage Came to Be Used by the Phage Group," Journal of the History of Biology, vol. 26, no. 2. 1993. pgs. 265–267.

material used and to nonspecific effects of the broth filtrates." Not long after this review was published, the number of phage therapy papers cited in the Index Medicus fell from 30 to roughly 12 per year.

***

Meanwhile, new medical discoveries were percolating elsewhere. In late 1932, a team of scientists working at a large German chemical company strung together two unlikely compounds and found they could kill strep bacteria faster than anything they'd ever seen. The company was Bayer and the new drug was Sulfanilamide, a unique combination of sulfur and red dye that would usher in the age of synthetic superdrugs.

As compared to bacteriophages, sulfa drugs, as the class of medicines came to be known, produced results that were relatively uniform and simple to reproduce in the lab. Most importantly, they had a far broader spectrum of action than phages. Word of the drug spread slowly in Germany, then leaked to England and France and, finally, burst onto the U.S. scene in the mid-1930s when doctors used it to save the life of President Franklin Delano Roosevelt's son, FDR, Jr.

Just before Thanksgiving in 1936, the 22-year-old FDR, Jr. was admitted to Boston's Massachusetts General Hospital with a sinus infection. Before long, the bacteria formed an abscess under his right cheek and began eating away at his throat tissue. The patient started coughing up blood. Before World War II, such a fast-spreading strep infection was almost certainly a death sentence. But FDR, Jr.'s physician, a top otolaryngologist named George Loring Tobey, Jr., was one of the first American doctors to test Prontosil, the Bayer sulfa drug, and had already used it on some of his sickest patients. "It was still unproven in his view, still experimental," writes Thomas Hager in "The Demon Under the Microscope," his history of sulfa, "and it went against the grain to try an experimental drug on the president's son, but he was running out of options." Dr. Tobey told Mrs. Roosevelt about Prontosil. For a second opinion, she called Johns Hopkins University, the largest U.S. medical center that was using the drug, and spoke with a physician who assured her of Prontosil's mild side effects. She gave her son's medical team the go-ahead to start treatment.

Three weeks into his hospital stay, FDR, Jr. received his first injection of the clear, blood-red medicine, and his first dose of sulfa pills, and he soon began to feel better. His skin temporarily turned tomato red, a common side-effect, but his fever came down, the swelling in his throat subsided and the abscess under his cheek dissolved. He was soon discharged and, next June, married his fiancée, Ethel du Pont, who had sat steadfastly by his bedside.

News of FDR, Jr.'s fast recovery leaked to the press. The New York Times broke the story on Dec. 17 with the front-page headline, "Young Roosevelt Saved by New Drug." The story went on to hail the medicine as "the most outstanding contribution of a decade to the conquest of infectious disease."[3] Sulfa drugs had

---

[3] The Demon Under the Microscope, by Thomas Hager 2006 Harmony Books, New York (pg. 189).

serious side effects, including vomiting and kidney damage, but their relatively simple administration and rapid results showed that synthetic antibiotics could have a powerful, curative effect. This was reflected in the explosion of literature devoted to them beginning in 1937 – instantly topping the space ever given to bacteriophage therapy.

Meanwhile, phage therapy papers continued at enough of a trickle to prompt JAMA to publish a follow-up to its first review in 1941. This one was even more unfavorable. The authors summarized numerous studies that appeared to show positive results from phage therapy but concluded that these were conducted either on too small a population or without adequate controls. After stating even more emphatically, and just as incorrectly, that "the nature of bacteriophage is no longer in question. It is a protein of high molecular weight and appears to be formed from a precursor originating within the bacterium,"[4] the authors went on to conclude that phage was of very little therapeutic value: "Modern chemotherapeutic approaches to the treatment of a variety of conditions for which phage has been recommended (cystitis, pyelitis, gonorrhea, certain bacteremias) offer more chances of success than does phage.[5]

The year following the JAMA review, for the first time in 2 decades, no studies were published on phage therapy in any major medical journal. All focus seemed to have shifted to synthetic drugs.

***

Even as sulfa continued to make headlines around the world for its potency, a group of researchers in England were working on an even more promising compound. In 1939 Ernst Chain, a young German-Jewish biochemist at Oxford rediscovered a decade-old paper by Alexander Fleming, a professor of bacteriology at St. Mary's Hospital in London. The paper was based on a chance observation that has passed into legend: Fleming noticed a colony of mold growing in a Petri dish in his lab. Upon closer inspection, he saw that the fluffy white mass, which had formed on a colony of staph bacteria, was surrounded by a halo of transparent, dead cells – much like the glassy areas first described by d'Herelle and Twort. Something in the mold had killed off the bacteria.

As Fleming pursued his discovery, he learned that the mold was a strain of *penicillium* (scientists later learned it was *penicillium notatum*) and that it could

---

[4] In fact, the status of phage as a live organism had been proven 1 year earlier, in 1940, by the German scientist H. Ruska, who had confirmed d'Herelle's theory by peering into one of the world's first electron microscopes.

[5] Albert Paul Krueger and E. Jane Scribner, "The Bacteriophage: Its Nature and Its Therapeutic Use," Journal of the American Medical Association, May 10, 1941, vol. 116, no. 19, pg. 2160–2167 and 2269–2277.

destroy many of the day's fiercest germs in the test tube, including strep, staph, and the bacteria that cause such illnesses as pneumonia and diphtheria. He also performed toxicity tests by injecting the mold extract into a mouse and a rabbit and irrigating large infected surfaces in volunteers. It produced no serious side-effects. "The toxicity to animals of powerfully antibacterial mould broth filtrates appears to be very low," he concluded in his 1929 paper on his discovery.[6]

Fleming saw right away that his discovery held tremendous promise as a drug. But he was unable to extract enough of its active ingredient, which he named penicillin, to identify it and test it in patients on a wide scale. Fleming worked on penicillin in fits and starts until finally abandoning it in frustration in 1935. Writes Eric Lax in his history of the drug, "[Fleming] is rather like a man who, say, discovered sparkling stuff in the water at Sutter's Mill in California in 1848, saw that it looked like gold," but was unable to mine it.[7]

Chain, who was working under Oxford's Chair of Pathology, Howard Florey, took an immediate interest in the substance and began experimenting on it. He didn't initially imagine that penicillin could turn out to be the wonder drug of the century. His aims were more modest: he was hoping that the mold could help his team learn more about the cell walls of pathogenic bacteria and how to break them down.

But, in the course of his team's research, they were able not only to extract penicillin and confirm Fleming's results but also to produce enough of it to test on animals and, eventually, humans. In May 1940, Florey and his coworkers infected eight white mice with strep and injected half of them with various doses of penicillin. The other four were left untreated as controls. Less than 24-h later, the results were in: all the untreated mice had died, and all the treated mice survived. As the group repeated its trials on larger and larger groups of mice, the results were enough to convince even the skeptical Florey. "It looks like a miracle," he exclaimed to a co-worker.

Just as Florey and Chain were getting their most promising results, it looked increasingly likely that the Second World War would come to British soil. It eventually did, in the blitz of London and other major cities that lasted from Sept. 7 until Oct. 31, when Hitler turned his attention to Russia. From that time on, it became impossible for the Oxford team to find enough supplies to produce penicillin in the quantities required to run large patient trials. "By April 1941," writes Lax, "the demands of the war and the bureaucracy of defense had combined to complicate the manufacturing of anything in England. Permits were needed for raw materials, and

---

[6] "On the antibacterial action of cultures of a Penicillium, with special reference to their use in the isolation of *B. influenzae*," Alexander Fleming, pg. 226–236. British Journal of Experimental Pathology, vol. 10, 1929.

[7] "The Mold in Dr. Florey's Coat," by Eric Lax, Henry Holt and Company, New York, 2005, pg. 3.

replacement parts of machinery (when available) required the processing of a sheaf of forms that moved slowly from one government office to another. This, coupled with only a small amount of clinical data on penicillin's effectiveness in humans, led to the inescapable conclusion that no English pharmaceutical company would help develop a more effective method of producing it."[8]

So, in the summer of 1941, Florey and Norman Heatley, a key member of his team who specialized in growing penicillin mold, headed to the United States to drum up partnerships with American pharmaceutical companies. With Florey's good connections, he soon had the government's Committee on Medical Research on his side. The Committee, part of the Office of Scientific Research and Development, helped organize a massive British-American effort that involved thousands of people and some 35 institutions, including the American pharmaceutical companies Merck, E. R. Squibb and Pfizer. Together, the team launched clinical trials of penicillin that would send Florey traveling to such far off places as North Africa and the Soviet Union to test penicillin in the field. Troops suffering from such fatal or hard-to-treat conditions as gangrene and gonorrhea were cured in a matter of days. In 1944, penicillin was finally introduced into civilian hospitals and earned Fleming, Florey and Chain the Nobel Prize the following year. They had ushered in the era of antibiotics.

***

Though many assume that the introduction of penicillin put an immediate end to phage therapy, that wasn't exactly the case. As soon as the sulfa drugs and penicillin were developed, scientists learned that bacteria could quickly adapt to them and spin off resistant strains. In response, a researcher working in Alexander Fleming's lab at St. Mary's Hospital in London tested bacteriophage and penicillin together in vitro. The results, published in 1945, showed that a strain of staph bacteria that was moderately resistant to both penicillin and phage became much more susceptible to destruction when attacked by both.[9] One year later, Ward MacNeal of Columbia University confirmed these results and tested the two agents together in patients suffering from ailments like bacterial endocarditis, peritonitis, bacterimia, osteomyelitis and boils caused by staph and step bacteria. "The conjoined action of … bacteriophage and penicillin has been remarkably promising in our experience during the last few years," he wrote.[10] His study was never followed up in the United States, although doctors in Poland, Russia and the Republic of Georgia continue to use bacteriophage alone and in combination with antibiotics.

---

[8] Lax, pg. 157.

[9] F. Himmelweit, "Combined Action of Penicillin and Bacteriophage on Staphylococci," The Lancet, July 28, 1945, pg. 104–105.

[10] Ward J. MacNeal, Louise Filak, and Anna Blevins, "Conjoined Action of Penicillin and Bacteriophages," Journal of Laboratory and Clinical Medicine," vol. 31, Sept. 1946, p. 974.

Not only was resistance a problem, but there were certain bacterial ailments for which even penicillin would prove ineffective. Chief among these was dysentery, the first disease against which d'Herelle had tested his phages and which remained a scourge of battlefield troops fighting in World War II. Allied as well as Axis commanders were eager to rid their ranks of this disease, which spreads through contaminated food and water, and which could land soldiers in the hospital for as long as 2 weeks with severe cases of bloody diarrhea.[11] In response, the U.S. government's Office of Scientific Research and Development began soliciting proposals from scientists aimed at finding a rapid cure. Rene Dubos, the prominent French-born bacteriologist suggested using phages. "We have obtained unequivocal evidence that bacteriophage therapy completely controls the development of certain types of experimental infection with dysentery bacilli in white mice," he wrote in a letter to the OSRD in July 1942. "The findings with reference to bacteriophage might make it desirable to consider plans for a more thorough study of the use of this agent in prophylaxis and therapy."[12] Working at the Rockefeller Institute for Medical Research and at Harvard, Dubos had shown that phages, when injected into white mice via stomach tube or mixed into their drinking water, could prevent them from developing dysentery. Out of eight mice that had been infected with dysentery and treated simultaneously with phages, six survived, as compared to zero in a control group of eight mice. Dubos also showed, in the same experiment, that the phages protected the mice by multiplying in their bloodstream and in their brains.

Initially, the government agreed to fund Dubos's work. But in 1944 it explicitly cut off funding for the part of his research that involved phages. "It seems doubtful to the Committee that this is a profitable undertaking," it concluded.[13] The rejection letter cited a paper published in early 1944 by J. S. K. Boyd, a colonel in Great Britain's Royal Army Medical Corps in which he concluded that "specific bacteriophage administered orally has no prophylactic action in bacillary dysentery, is incapable of aborting an attack of the disease, and has no dramatic effect, if indeed it has any action at all, in modifying the severity and duration of an attack."[14] He based his conclusion on a controlled study involving hundreds of German POWs, where he'd found little difference between troops who had been given phages before or during the course of their illness and those who had not. With the end of World

---

[11] The German Army in Africa used phages as a standard treatment for bacillary dysentery, employing a preparation by Bayer called Ruhr-Bakteriophagen, which was elegantly packaged in brown glass bottles with rubber stoppers. Ref. is Boyd and Portnoy paper, cited below.

[12] Dubos to Dr. A. N. Richards.

[13] Letter from to Dr. A. N. Richards, Chairman of the Committee on Medical Research, OSRD from Philip S. Owen, M.D. of the National Research Council, May 12, 1944.

[14] "Bacteriophage Therapy in Bacillary Dysentery," Transaction of the Royal Society of Tropical Medicine and Hygiene, vol. XXXVII, No. 4, February 1944, J. S. K. Boyd and B. Portnoy, pg. 260.

War II near at hand, Boyd's was the last word on the effectiveness of bacteriophages on the battlefield.

During the 1940s, bacteriophage labs shut down as physicians moved on to pursue other interests. Ward MacNeal died in 1946, apparently leaving no heirs to his phage therapy practice. E. W. Schultz shut down his lab sometime during World War II, possibly frustrated by the lack of reliable data coming in from doctors. In a speech he gave in 1932,[15] he complained that "the analysis of these reports has been far from a simple and satisfying task" and noted his clients' consistent failure to include basic information in their case reports, such as a complete diagnosis, a clinical history, dosage and method of administration. He turned instead to research on poliomyelitis. When he died in 1971, not one of his obituaries mentioned his contribution to phage therapy. The field had been all but forgotten in the West.

[15] E. W. Schultz, "Bacteriophage as a therapeutic agent in genito-urinary infections," California and Western Medicine, vol. 36, no. 1, pg. 36.

Chapter 4
Naked Genes

American medicine may have abandoned phages by the end of World War II, but the viruses were about to land a starring role in the budding field of molecular biology. A group of now-legendary scientists plucked phages from near-obscurity and used them to peer into the fundamental building block of life: the gene. In doing so, they unraveled many of the mysteries that had long confounded phage researchers: whether or not phages are living organisms, how, if at all, they destroy bacteria and, on a more basic level, how they look.

The story begins in the late 1920s, with two future Nobel laureates struggling to find their respective paths in science. Salvador Luria, an Italian-Jewish medical student living in Turin and Max Delbruck, an astronomy student living in Berlin, would not meet for several years, but their interests were already beginning to converge.

Luria, feeling lukewarm about medicine, began gravitating toward research science instead. "Those were the years," wrote Luria in his memoirs, "1929–1930, when the new, revolutionary ideas of physics – quantum theory, wave mechanics – were filtering down to undergraduates, at least to the best ones, and with them, names like Bohr, Heisenberg, Schrodinger, and especially Enrico Fermi, the already legendary young star of Italian physics." A friend of Luria's was studying under Fermi at the University of Rome and arranged for Luria to take classes there as well. "That year among physicists proved to be the critical turning point in my life," he wrote.

Meanwhile, Max Delbruck, Luria's senior by 6 years, was falling in love with physics as well. For him, it was hard to avoid: Germany in the 1920s was at the epicenter of a series of revolutionary discoveries, including the development of quantum theory, which states that energy is radiated through fixed, elemental units, and of Niels Bohr's complementarity principal, which allowed particles to be described both as units and as waves spreading through space. In 1926, Delbruck came to the prestigious Gottingen University for 3 years to study astronomy but quickly became caught up in the excitement surrounding physics and switched specialties. He earned his doctorate in theoretical physics under future Nobel winner Max Born.

But physics, too, proved a mere stepping stone for both Luria and Delbruck. For Delbruck, another turning point came in 1932, as he was pursuing his first Rockefeller Fellowship in Copenhagen, attending sessions and seminars at the

A. Kuchment, *The Forgotten Cure: The Past and Future of Phage Therapy*,
DOI 10.1007/978-1-4614-0251-0_4, © Springer Science+Business Media, LLC 2012

Institute of Theoretical Physics under the leadership of Bohr. In August of that year, Bohr delivered a lecture called "Light and Life" in which he outlined how his complementarity principal might apply to biology. One of the ideas behind complementarity is that one cannot observe a particle without altering its behavior; even the seemingly benign ray of light that allows a scientist to observe matter, for instance, consists of photons that exert pressure on the particle. In applying his law to biology, Bohr suggested that one might never truly be able to understand a living organism. "Thus, we should doubtless kill an animal if we tried to carry the investigation of its organs so far that we could describe the role played by single atoms in vital functions. In every experiment on living organisms, there must remain an uncertainty as regards the physical conditions to which they are subjected, and the idea suggests itself that the minimal freedom we must allow the organism in this respect is just large enough to permit it, so to say, to hide its ultimate secrets from us."[1] The notion that life, which seemed so hopelessly complex, might be boiled down to simple, universal laws, intrigued Delbruck, and he spoke of Bohr's lecture throughout his life as having provided the sole motivation for his work.[2]

The next step in Delbruck's scientific self-discovery came later that same year when he moved back to Berlin and became an assistant to Lise Meitner at the Kaiser Wilhelm Institute for Chemistry. (At the time, Meitner was on her way to the discovery of the fission of uranium, the process that was later used in the explosion of the atomic bomb). The Kaiser Wilhelm Institutes, among the most respected centers of science in the world, brought Delbruck into close proximity with scholars across a wide spectrum of fields. He soon met the Russian geneticist Nicolai Timofeev-Ressovsky, who, with Herman J. Muller, was trying to provide definite proof that x-rays caused mutations in genes. Because Delbruck was searching for a way of applying the laws of physics to biology, they formed an immediate connection. "Two or three times a week we met," recalled Karl Zimmer, one of their collaborators, "mostly in Timofeeff-Ressovsky's home in Berlin, where we talked, usually for 10 h or more without any break, taking some food during the session. There is no way of judging who learned most by this exchange of ideas, knowledge and experience, but it is a fact that after some months Delbruck was so deeply interested in quantitative biology, and particularly genetics, that he stayed in this field permanently." Timofeeff-Ressovsky, Delbruck and Zimmer undertook a revolutionary experiment: the trio aimed x-rays at the genes of Drosophila fruit flies and demonstrated that the genes mutated proportionally to the given dose of radiation. In their resulting 1935 paper, "About the Nature of Gene Mutation and Gene Structure," the trio postulated that genes could prove to be "the ultimate units of life." They had succeeded in providing further evidence that genes were tangible molecules that followed the laws of quantum physics – their behavior could indeed be measured and predicted.

Published in an obscure journal, the paper might easily have been forgotten if the great Austrian physicist Erwin Schrodinger had not come upon it and included it in his influential 1947 book, "What Is Life?" In the slim volume, which contained

---

[1] "Thinking About Science," 82.

[2] "Thinking About Science," 82.

a chapter called "Delbruck's Model Discussed," Schrodinger used the trio's experiment to underline the notion that genes are essentially "stable" – they require a high dose of energy, or heat, in order to change or mutate. How else could a feature like the Habsburg lip be passed down through so many generations? In considering the physical nature of the gene, Schrodinger concluded that "we may safely assert that there is no alternative to the molecular explanation of the hereditary substance. The physical aspect leaves no other possibility to account for its permanence. If the Delbruck picture should fail, we would have to give up further attempts."[3] Schrodinger went on to presciently speculate that each gene was made up of a small number of units – now known as base pairs – that can be rearranged in different combinations. "Indeed, the number of atoms in such a structure need not be very large to produce an almost unlimited number of possible arrangements. For illustration, think of the Morse code. The two different signs of dot and dash in well-ordered groups of not more than four allow of 30 different specifications."[4]

Back in mid-1930s Rome, Luria was gradually coming to the conclusion that his interest in physics "was bound to remain amateurish rather than professional." And, like Delbruck, began trying to think of a field that could bridge biology and physics. He found it in radiation biology, the field that Delbruck was helping to pioneer. One day, one of Luria's professors, an omnivorous reader, handed Luria the three man paper by Delbruck, Ressovsky and Zimmer. "These papers seemed to me, however ignorant of genetics I then was, to open the way to the Holy Grail of biophysics – and there and then, I believe, I swore to myself that I would be a knight of that Grail," wrote Luria. After reading Delbruck's papers, Luria began trying to think of a biological object on which to test Delbruck's ideas on the gene, and that's when he discovered bacteriophages.

He heard about them from an acquaintance he met on the Rome trolley, a professor of virology who was using phages to test the Tiber river for dysentery. He came to his friend's lab to learn more about the microbes. "Between bacteriophage and myself it was love at first sight," wrote Luria. "All week I played with test tubes and Petri dishes, devising new ways of growing and counting bacteriophage, even using some of the simple statistical methods I had learned in physics – making all possible mistakes but for the first time in my life getting excited about research."

Delbruck was gravitating toward bacteriophages at almost the same time. As early as 1922, Muller, who would work closely with Delbruck and Timofeev-Ressovsky in Berlin a decade later, suggested that bacteriophages, because they appeared to be the smallest observable entities capable of reproduction, might be naked genes. "If these d'Herelle bodies [bacteriophages] were really genes, fundamentally like our chromosome genes, they would give us an utterly new angle from which to attack the gene problem," he wrote. "… It would be very rash to call these bodies genes, and yet at present we must confess that there is no distinction known

[3] Schrodinger, Erwin, "What Is Life?" Cambridge: At the University Press, 1947, pg. 57.
[4] "What Is Life?" p. 64.

between the genes and them. Hence we cannot categorically deny that perhaps we may be able to grind genes in a mortar and cook them in a beaker after all."[5]

Delbruck's final decision to study bacteriophages came in the late 1930s, after he came to the California Institute of Technology (Caltech) on his second Rockefeller Foundation Fellowship. At Caltech, he met Emory Ellis, a postdoctoral research assistant who had been using phages to shed light on the role of viruses in cancer. "I had vaguely heard about viruses and bacteriophages … before I left Germany," Delbruck recalled in 1978. "I had sort of the vaguest notion that viruses might be an interesting experimental object for a study of reproduction at a basic level."[6] Delbruck and Ellis began studying the viruses together and, in 1939, published a joint paper on phage growth. Word that Delbruck was working with phages trickled back to Luria in Rome: "As in a troubadour romance," wrote Luria, "the love triangle among Delbruck, bacteriophage and me had come into being at 4000 miles' distance and without mutual awareness of each other."

Delbruck and Luria finally met 2 days before New Year's Day of 1941 at an American Physical Society convention in Philadelphia. Both had taken up residence in the United States to escape their countries' fascist regimes. Luria, who was Jewish, was more directly effected by the political changes. After Mussolini proclaimed his notorious "Racial Manifesto" in July 1938, aligning Italy with Nazi Germany and excluding Jews from the Italian race, Luria's father lost most of his accounting business, his brother lost his job and Luria's hard-won Italian government scholarship, which he was going to use to travel to the United States to work with Delbruck, was revoked. Luria traveled to still-unoccupied Paris and, when the German army crossed into France in June 1940, he fled to Marseilles by bicycle – a month-long journey that took him through Limoges, Bordeaux and Toulouse in the summer heat. In Marseilles, he wrangled a visa, came to New York and – thanks to Fermi who, with his Jewish wife, had already escaped to the United States – secured a fellowship at Columbia University working with the physicist Frank Exner.

Delbruck wasn't Jewish, but his career still suffered under the Nazis. "Max tried to follow orders," write his biographers, Ernst Peter Fisher and Carol Lipson. "Since he wanted to become a lecturer at the university, he had to fulfill the Nazis' condition: to gain approval politically. To do so, he had to attend a Dozentenakademie, as it was called; one can best describe it as an indoctrination camp. About 30 people would gather in discussion groups to hear lectures on the new politics and the new state. After 3 weeks of "free" discussions, the organizer determined whether a participant was politically mature enough to lecture at the university." Apparently, Delbruck never passed muster.[7] Along with dozens of other prominent intellectuals,

[5] H. J. Muller, "Variation Due to Change in the Individual Gene," Amer. Nat., 56 (1922), cited in Summers, "How Bacteriophage Came to Be Used by the Phage Group," pg. 262.

[6] Delbruck, Oral history 1978, pg. 63, Delbruck Papers, Caltech Archives, cited in Summers, William C., "How Bacteriophage Came to Be Used by the Phage Group."

[7] "Thinking about Science: Max Delbruck and the Origins of Molecular Biology," Ernst Peter Fischer and Carol Lipson. W. W. Norton & Company, New York. 1988, pg. 73.

Delbruck left his homeland. The mass exodus effectively lobotomized a country that was once a formidable leader in science.

Shortly after they met, Delbruck, who was by then teaching physics at Vanderbilt University, visited Luria in New York, and the two spent New Year's Day conducting phage experiments. "From the start, Delbruck struck me as a dominant personality," wrote Luria. "Tall, and looking even taller because of his extreme thinness, moving and speaking sparingly and softly but with great precision, he conveyed the impression that whatever he said had been carefully thought out. His seriousness was occasionally broken by sparks of amusement, often produced by unexpected contrasts, and especially at the expense of someone's pretentiousness. His humor was usually gentle but could be deflating, although never cruel." Luria and Delbruck had contrasting personalities. Luria was quiet, serious and a romantic. Delbruck was a charismatic leader: a disciplined worker who also liked to have fun in the form of parties, cookouts, tennis, and camping trips. "The difference in our personalities constituted the attraction at the base of our friendship," wrote Luria.

It was during their first meetings that Luria and Delbruck planted the seeds of what would later become known as The Phage Group, a loose collection of scientists in diverse fields who believed that bacteriophages constituted the biological equivalent of hydrogen atoms: the simplest and smallest life forms. Between 1945 and the late 1960s, an informal school grew up around the Phage Group that was centered on a summer course taught at the Cold Spring Harbor Laboratory on New York's Long Island. "The purpose of this course was frankly missionary," wrote one of its famous alumni, the physical chemist and historian of science Gunther Stent. Its mission was to indoctrinate chemists, biologists and physicists in the revolutionary discipline that would mark the birth of molecular biology: the use of the bacteriophages as a tool for studying genes. The course attracted the brightest scientific minds of the day. James Watson was a Delbruck protégé, as were Renato Dulbecco, who in 1975 won the Nobel Prize for his work on tumor viruses. Established investigators, too, were drawn to the course, including the father of the nuclear chain reaction, Leo Szilard, and the atomic physicist Philip Morrison.

Beyond hard science, the group's summer course provided an idyllic retreat for students and faculty who lived together on the bucolic laboratory campus on the shore of Long Island Sound. On weekends and during breaks, students went sailing or canoeing, at night they hobnobbed with the best minds in their profession – or went skinny-dipping. In his essay in Phage and the Origins of Molecular Biology, Watson recalls a jocular atmosphere where top intellectuals let loose and behaved like kids at summer camp: they staged attacks with toy machine guns, let the air out of each other's tires and poured buckets of water over students' beds.[8]

One classic event was the staging of Shakespeare's "A Midsummer Night's Dream" on the porch of one of the lab buildings. Max Delbruck, dressed in a toga, sandals, and a diadem of drooping leaves, played Theseus, and his wife, Manny, played Hippolita. The rest of the cast was made up of lab staff, maintenance men, neighbors, anyone

---

[8] Phage and the Origins of Molecular Biology, Watson, 244.

they could round up. Writing in 1992, John Cairns, one-time director of CSHL, recalled how his mentor performed Theseus's final speech from the play:

> I remember that [Delbruck] would intone this with a wonderfully rolled R in the word "iron." It comes at the end of the play, when most of the actors have departed and the audience is in a high state of entrancement. Add to this the steamy heat of a Long Island summer and the incessant call of the cicadas, and imagine that you have just come out of your air-conditioned laboratory into the immediacy and heaviness of night. Then perhaps you will begin to see what it was like, in those far off days, when molecular biology seemed innocent and incorruptible.[9]

These kinds of diversions helped foster a spirit of collaboration that was the group's hallmark and that helped it move, in short order, from one exciting breakthrough to the next. One of the earliest was the discovery that phages had complex structures – they weren't simple spheres as many scientists had assumed. The development of the electron microscope in Germany in 1940 enabled phage workers to glimpse their subject for the first time. Luria brought a sample of phage to the RCA Manufacturing Company in Camden, NJ in March 1942 to have the electron microscopist Thomas Anderson photograph it. They observed "tadpole-shaped particles" that readily attached themselves to bacterial cells. When they sent their images to other phage workers, many were stunned. "Mein Gott! They've got tails!" exclaimed Hershey's teacher, the longtime phage scientist J.J. Bronfenbrenner of Washington University in St. Louis.[10]

One problem that confounded the group was how phages penetrate bacteria and replicate inside them. When viewed through a microscope, it appeared that the bacteriophages remained on the outside of the bacterial cell. How, then, did they get inside – if at all? By the early 1950s, scientists knew that phages consisted of DNA, which they stored in their "heads," and protein, which made up their outer shell, including the "tail." But it was still not known whether the protein or the DNA made up the genetic material. "If the particles do penetrate the bacterial ghosts they must become unrecognizable," Anderson theorized at the time. "However, it also seemed possible they do not penetrate; and indeed, one could see empty-headed phage ghosts on the bacterial surface. I remember in the summer 1950 or 1951 hanging over the slide projector table with Hershey ... at the Cold Spring Harbor Laboratory, discussing the wildly comical possibility that only the viral DNA finds its way into the host cell. ... In the hands of Hershey and Chase (1952) this joke proved to be not only ridiculous but true."

In 1952, Hershey and collaborator Martha Chase tagged phage DNA and phage protein with two different kinds of radioactive tracers. They then combined the phages with *E. coli* bacteria, spun the mixture in a blender and then whirled it in a centrifuge to separate the heavier infected bacteria from the lighter phage particles.

---

[9] John Cairns, Preface to the Expanded Edition, "Phage and the Origins of Molecular Biology." 1992 viii.

[10] "Electron Microscopy of Phages," by Thomas F. Anderson pg. 64–65, in Phage and the Origins of Molecular Biology.

Hershey and Chase then tested the bacteria and the phage particles for radioactivity. The results: most of the phage DNA had entered the bacterial cells and most of the phage protein had remained outside. This was proof that DNA, and not protein, was the genetic material, and this discovery would pave the way for James Watson and Francis Crick's breakthrough resolution of the structure of DNA the following year, which led to our current understanding of how genes replicate.[11]

The discovery had other consequences as well. If phages contain DNA and DNA is genetic material, then phages were indeed living organisms and not enzymes, as d'Herelle's opponents had insisted – although, unlike bacteria and higher organisms, viruses are incapable of replicating on their own without invading a host.

With Hershey and Chase's discovery, the Phage Group, for the first time, had a nearly complete picture of how phages attack and destroy bacteria. It's an intricately choreographed process worthy of George Balanchine. The phage is essentially a "tiny disposable protein syringe loaded with viral DNA," wrote Anderson.[12] It attaches to a bacterial cell and shoots its DNA inside, where the particles replicate to make dozens of daughter phages. In the space of 10–15 min, as many as 1000 new phages can form inside the bacterium, eventually bursting out of the cell and swimming off in search of more targets.

In 1969, Hershey, Luria and Delbruck won a joint Nobel Prize for their discovery of the replication mechanism and genetic structure of bacteriophages. "At first the scientific community in general had struck a reserved attitude to bacteriophage research," said Sven Gad, a professor at the Royal Caroline Institute, who delivered the Nobel presentation speech. "It was considered to be of interest as a curiosity but of little importance to biology in general. Gradually this attitude has changed. It is now clearly evident that in principle the same mechanisms regulate the activities of bacteriophages, micro-organisms and more complex cellular systems. Therefore, Delbruck, Hershey, and Luria must in fact be regarded as the original founders of the modern science of molecular biology." He added that Delbruck had "transformed bacteriophage research from vague empiricism to an exact science."

Though Gad, in his comments, appears to dismiss d'Herelle's work, the self-taught scientist's observations are all the more remarkable in light of the Phage Group's discoveries. Working with only a light microscope and without knowledge of DNA, d'Herelle was able to correctly guess how bacteriophages penetrated and destroyed their prey. Upon adding phages to the bacteria, he reported in a note written in 1921, he saw an increase in the number of bright microscopic "points" inside the bacterial cell. After a while, more points appeared and the cell puffed out into a spherical shape.

---

[11] Hershey and Chase's results confirmed those from a 1943 experiment by Oswald Avery and colleagues at the Rockefeller Institute that found that DNA, not protein, appeared to be "the fundamental unit of the transforming principle" of bacteria that cause pneumonia. Paper reference: Oswald T. Avery, et al., "Studies on the Chemical Nature of the Substance Inducing Transformation of Pneumococcal Types: Induction of Transformation by a Deoxyribonucleic Acid Fraction Isolated from Pneumococcus Type III, Journal of Experimental Medicine 79 (Feb. 1, 1944).

[12] Ibid., 76.

Finally, after the germs were destroyed, the points dispersed throughout the medium.[13] "The image of the process of multiplication of bacteriophage given by d'Herelle (1926) was reasonable, and in accord with the experimental observations, but seemed to attract more opponents than supporters," wrote Emory Ellis in 1966. "Considering that the powerful present-day experimental tools (electron microscope, analytical centrifugation, and radioactive tracer techniques, among others) were not available to d'Herelle, his description of the growth process struck remarkably close to the picture we have today."[14]

And yet the Phage Group inadvertently helped shut the door to further progress in phage therapy. The scientists who surrounded Delbruck were concerned with one central problem: how higher organisms reproduce. Any role that phages might once have played in curing human disease was rarely thought of and dismissed as a failed effort. In his 1966 introduction to "Phage and the Origins of Molecular Biology," a *festschrift* dedicated to Max Delbruck on his 60th birthday, Stent summarized the virus's transition from hospital bedside to lab bench. "Bacteriophages … came to play a glamorous role in the bacteriology of the 1920s. By the middle of the 1930s, however, this glamour had begun to tarnish, since the widely propagandized control of bacterial diseases by means of bacteriophages had failed to materialize."[15] Three years earlier, Stent had written an influential textbook that would shape the attitudes of generations of scientists toward bacteriophage therapy. "Just why bacteriophages, so virulent in their antibacterial action in vitro, proved to be so impotent in vivo has never been adequately explained," he wrote in Molecular Biology of Bacterial Viruses. "Possibly the immediate antibody response of the patient against the phage protein upon hypodermic injection, the sensitivity of the phage to inactivation by gastric juices upon oral administration, and the facility with which (as we shall see presently) bacteria acquire immunity or sport resistance against phages, all militated against the success of phage therapy."[16] Indeed, one of Delbruck and Luria's most famous experiments, the fluctuation test that they performed in 1943, showed that bacteria spontaneously mutate to become resistant to phages. A clear tube containing bacteria that has been destroyed, or lysed, by phage will cloud over again in a matter of hours as new, resistant strains of bacteria begin to reproduce. Their results formed the cornerstone of bacterial genetics but also left the lingering doubt that bacteriophage could really conquer bacteria.

And, though the Phage Group's contribution to the fields of genetics and virology was profound, the scientists based their results on studying a very small number of viruses. In 1944, Delbruck made the decision to focus his team's work on

---

[13] Summers, "Felix d'Herelle", pg. 67.

[14] Emory L. Ellis, "Bacteriophage: One-Step Growth," in "Phage and the Origins of Molecular Biology," p. 57.

[15] John Cairns, Gunther S. Stent, James D. Watson eds., "Phage and the Origins of Molecular Biology," Cold Spring Harbor Laboratory of Quantitative Biology, Long Island, NY, 1966, pg. 5.

[16] Stent, Gunther "Molecular Biology of Bacterial Viruses," W. H. Freedman and Col, San Francisco, 1963, p. 8.

just seven phages. This was so that each researcher would build on every other researcher's work – he wanted results to be easily comparable across the spectrum. The Phage Group chose seven closely related *E. coli* phages – nicknamed the seven dwarfs – on which to focus their studies. They were named T1 through T7, with the "t" standing for "type." As a result, biologists today know almost nothing about phage diversity – how many different types of phages there are in the wild, and how representative the seven dwarfs are of the rest. "The knowledge is a mile deep and an inch wide," says Ry Young, a longtime phage biologist at Texas A&M University.

In the early 1950s, having decided that bacteriophages were "in good hands," Delbruck shifted his focus to the study of the higher nervous system.[17] The course continued to attract hundreds of scientists through the 1960s, after which time many moved on to other organisms, like yeast or back to *Drosophila*. As technology advanced to allow scientists to work more easily with human and animal cells, the field of phage biology virtually died out in the 1970s. By one count, applications to the National Institutes of Health for grants to study phage biology numbered 280 in 1972; 30 years later, that number had fallen to 10.[18]

---

[17] Phage and the Origins of Molecular Biology, Stent, 7.

[18] Interview with Ryland Young, Texas A&M. Fall, 2002.

Chapter 5
'They're Not a Panacea:' Phage Therapy in the Soviet Union and Georgia

Since Eliava and d'Herelle introduced phage therapy to the USSR in the 1930s, its popularity has waxed and waned, but, unlike in the United States, it never fell out of use. Thanks to the duo's earlier promotional efforts and speaking tours, phage therapy labs sprang up across the empire in the 1940s in such cities as Moscow; Gorky and Alma-Ata, Kazakhstan. Physicians administered bacteriophages orally, in liquid or tablet form; topically, for skin, eye and ear infections; by aerosol for respiratory infections and, on rare occasions, intravenously to treat blood infections.[1]

One of phage therapy's most active proponents at that time was Alexander Tsulukidze, professor of general surgery at the Tbilisi Medical Institute and the first to implement the use of phages in preventing and treating post-operative infections. He published 14 papers on the subject between 1935 and 1957.[2] While Eliava was still alive, in 1936, Tsulukidze contributed a paper to a special edition of the French journal La Médecine that had been dedicated to phage therapy. In the paper, he described the treatment of 47 patients with intestinal perforations, a complication of typhoid that was almost always fatal in the pre-antibiotic era. Even with surgery to repair the perforation, Tsulukidze cited the mortality rate at his institution as 89% within 3 days of the operation. He divided his patients into two groups: 27 received only supportive care post-surgery while 20 patients received phage therapy in addition to the standard regimen. In the control group, the mortality rate was 89% (24 out of 27 patients died) while among the group that received phages, only 35% (7 out of 20 patients) perished. (The 1941 JAMA review concluded that this study was performed on too small a population for the results to be of significance.)

In the same journal, Tbilisi physician Charles Mikeladze presented evidence that phages, given orally and injected by IV, were effective against advanced cases of

---

[1] Bacteriophages: Biology and Application, Elizabeth Kutter and Alexander Sulakvelidze, eds. Chapter: Bacteriophage Therapy in Humans, by Alexander Sulakvelidze and Elizabeth Kutter, pg. 400 (CRC Press, Boca Raton, Florida, 2004).

[2] Summers, William C. "Felix d'Herelle and the Origins of Molecular Biology," pg. 163.

A. Kuchment, *The Forgotten Cure: The Past and Future of Phage Therapy*, DOI 10.1007/978-1-4614-0251-0_5, © Springer Science+Business Media, LLC 2012

typhoid fever. Eighty-five patients were split into two groups, one of which served as a control. Of the 64 control group patients who received standard care, which at that time probably consisted of fluids and supportive therapy, 15.6% died, as compared with 4.8% of those treated with bacteriophage. Patients in the control group also had a higher rate of complications, 56.2% as compared to 13% but a lower rate of relapse (4.5%, vs. 9.5% of those who survived).[3]

Once World War II broke out, physicians deployed phages to help prevent and treat infections on the battlefield. Magdalina Pokrovskaya, a physician who had worked with d'Herelle, conducted a trial on soldiers injured during the Soviet Union's invasion of Finland in late 1939. Her boss, Brigadier Zhuravlev, chairman of the Medical Directorate of the Red Army, wrote an introduction to her findings, which were compiled and published as a stand-alone book. The study used phages from the Eliava Institute as well as from Moscow's Metchnikov Institute and, in most cases, applied them by liberally spraying wounds with them. In other cases, phages were injected subcutaneously (beneath the skin) or, in severe cases of sepsis, were used intravenously. Though there were no control groups, the authors claimed that the patients' overall condition improved rapidly with phages and that their wounds healed quickly and were less prone to infection, especially when phages were administered within the first 24 h. Because of extensive studies done on phage therapy by physicians in France, the USSR and the United States, the paper concludes, "phage are being increasingly accepted into the medical arsenal of effective tools of modern surgery, and for this reason, every surgeon should be acquainted with the important attributes of this preparation."

As well as giving evidence of the use of phage therapy at the highest government levels, Dr. Pokrovskaya's report hints at a possible reason why phage therapy studies yielded such conflicting results: phages were incredibly complex to use and to store. Because modern techniques like spray-drying and freeze-drying had not yet been developed, phages had to be kept in a liquid broth. "Ampoules used for the preparation of bacteriophages must be made from the highest quality glass, a variety that does not produce alkali, or else over time the pH of the liquid will change and the phages will die," she wrote.[4, 5]

Just after World War II, American microbiologist Stuart Mudd visited the Soviet Union and observed phage therapy in use across Russia and Georgia. He described the country's medical and health care facilities in a paper published in 1946 in the journal Science and in the American Review of Soviet Medicine. Among other observations, Mudd reported that, in Moscow, the Central Institute of Epidemiology and Microbiology of the Ministry of Public Health produced bacteriophage, in

---

[3] Summers, 170.

[4] Summers, 171.

[5] Pokrovskaya, M.P.; Kaganova, L. C.; Morozenko, M. A.; Bulgakova, A. G; Skatchenko, E. E. *Letchenye Ran Bakteriofagom*. [The cleaning of wounds using bacteriophage], Chief Military-Medical Administration of the Red Army, Government Publishing House of Medical Literature 'Medgiz'. Moscow. 1942.

addition to a number of other vaccines and serums, "on a large scale." During the war, he noted that scientists scattered bacteriophage in soil, sewage and food and administered it prophylactically to soldiers as a means of preventing dysentery. "The director [of the phage production facility] Professor L. Yankelevich, said that statistics regarding bacteriophage were weak; he had observed its therapeutic use, however, and thought the results were good," Mudd reported. He also noted that the All-Union Institute of Biological Prophylaxis of Infections, also in Moscow, produced and distributed a bacteriophage preparation against diarrheal diseases and distributed it to children in summer camps and other programs every 10 days between May and October. " ... [G]roups of children had received the phage and control groups had received a physiologic salt solution, with a lower incidence in the protected than in the control groups. Statistical studies are promised for the American Review of Soviet Medicine." Mudd also visited the Eliava Institute in Tbilisi and watched the staff prepare liquid phage and phage tablets against dysentery.[6]

Antibiotics first reached the Soviet Union at around the same time they became available in the United States, but under very different circumstances. At the Tehran Conference in 1943, Winston Churchill, Joseph Stalin and Franklin Delano Roosevelt set the date for the 1944 invasion of Normandy, but they also agreed on another covert mission: sending four scientists, two from the United States and two from Great Britain, to Moscow to brief the Soviets on the latest Western advances. Howard Florey, who led the development of penicillin at Oxford University, was chosen to represent the UK, and he selected Gordon Sanders, an expert on penicillin production, to accompany him. The two arrived in January of 1944, bringing with them mould samples as well as "some of our best purified penicillin extract," recalled Sanders. They also briefed their hosts on about a dozen other scientific developments.[7] Led by Zinaida Ermolyeva, head of the Institute of Biochemistry in Moscow, and with funding from Great Britain, the USSR quickly began producing the antibiotic and, by D-Day, doctors were treating Allied soldiers with penicillin manufactured on Russian as well as on American soil.

Despite this promising start, antibiotic production in Russia would face a spotty future. With the end of World War II and the suspension of Western funding and support, the Soviets were forced to rely on their own scant resources, which included modest equipment and a lack of trained personnel. By some accounts, penicillin didn't go into wide-enough production to reach civilians until early 1949.[8]

The next major antibiotic, streptomycin, also appeared in the Soviet Union after a delay. Its story is filled with a similar amount of intrigue as that of penicillin. After

---

[6] Mudd, Stuart, "Recent Observations on Programs For Medicine and National Health in the USSR. Part Two. American Review of Soviet Medicine, vol. 5, 1947, pg. 74 and 75.

[7] Bickel, Lennard. "Rise Up To Life: A biography of Howard Walter Florey who made penicillin and gave it to the world." Charles Scribner's Sons. New York. 1973. Pg. 212.

[8] Savitskii, A. V. "Antibiotics – 25 Years of Science and Production in the USSR." Pharmaceutical Chemistry Journal. Vol. 1, No. 10. 1967. pg. 571.

it became available in the United States in 1944, the Soviet scientist Lina Stern obtained a sample through family connections and, within the next couple of years, became the first physician to successfully use it to treat tubercular meningitis. Despite her accomplishment, the drug didn't reach the Soviet mass market until the mid-1950s.[9] With the population clamoring for the wonderdrugs they had heard about, usually from relatives traveling abroad, the Soviet government turned to propoganda: just after the war, it ran a campaign urging Russians to use herbs and other natural remedies because they are "*nashi*," or "ours." The announcements tied the use of herbs to patriotism and painted pharmaceuticals, which were being developed and promoted by the West, as something foreign and suspect.[10, 11]

Russia's drug supply problems stemmed from a lack of funding. American pharmaceutical companies were reaping vast profits by developing, patenting and selling medicine. But in the USSR, all medical services – from pharmaceutical research and distribution to hospital care – were paid for and provided free of charge by the state. Yet the state failed to match its goal of universal health care with adequate resources. "Health care was a backwater area," says Mark Field, an expert on Soviet socialized medicine at Harvard University's Davis Center for Russian and Eurasian Studies. "It was of secondary importance to defense and to the industrial production of weapons." As a result, lifesaving drugs were of inferior quality and vulnerable to the same shortages as pencils and shoelaces. "Streptomycin, for example, will be on every shelf one week and then suddenly disappear the next," wrote one American medical observer who traveled through the country in the 1970s. While a typical American hospital stocked upwards of 60 antibiotics in the 70s, one Moscow hospital the observer visited stocked about eight, of which only four were available.[12]

In the United States, medical culture has long been rooted in science and technology, with patients and physicians relying on drugs that pass clinical trials under the supervision of the Food and Drug Administration. Alternative medicine – natural remedies that fall outside standard medical practice – didn't gain mainstream popularity until the 1990s, when insurance companies began covering herbal medicine, acupuncture and therapeutic massage. In the Soviet Union, however, alternative medicine has always been an essential part of treatment, perhaps because patients needed a fallback when standard medicines were unavailable. Drug stores routinely stocked herbs such as anise seeds for gas; dill seeds for heartburn and chamomile

---

[9] A. V. Savitskii, Antibiotics: 25 Years of Science and Production in the USSR, Pharmaceutical Chemistry Journal. Vol 1, No. 10, October, 1967.

[10] Dreifuss, Jean-Jacques; Tikhonov, Natalia. "The intersection of the personal and academic history: Lina Stern (1878–1968)." Paper presented at The Global and the Local: The History of Science and the Cultural Integration of Europe. Proceedings of the 2nd ICESHS (Cracow, Poland, September 6–9, 2006)/Ed. By M. Kokowski.

[11] Author interview with Mark Field, an expert on Soviet socialized medicine at Harvard University's Davis Center for Russian and Eurasian Studies, 8/19/07.

[12] "Inside Russian Medicine: An American Doctor's First-Hand Report," by Nicholas A. Petroff, Everest House Publishers, New York, pg. 128, 1981.

tea for ulcers.[13] One can see how phage therapy, which was always considered an alternative treatment, even in the Soviet Union, could more easily take root in a culture that trusts natural substances and where the government once attempted to program its citizens against the pharmaceuticals it often failed to provide.

A good example of how phages coexisted with antibiotics in the USSR came in the 1970s, during the construction of the Baikal-Amurskaya Railroad, or BAM. This 6,000-km railroad would link Eastern Siberia, with its vast mineral resources, including gold, diamonds and oil, with Russia's military industrial centers in Western Siberia and the Urals. When a medical team was dispatched to care for the workers, it quickly discovered a highly resistant outbreak of staph bacteria in the population, writes David Shrayer, a Soviet-trained physician who later immigrated to the United States. Some 80% of the staph strains were resistant to multiple antibiotics. Shrayer went to Siberia with the BAM Medical Council to care for patients, and he traveled along the route of the railroad. He described a particularly dire situation in a town called Nizhneangarsk:

> [The Nizhneangarsk Hospital] had 100 beds as well as surgical, therapeutic, obstetric and pediatric wards. Staff, visitors, and patients moved freely from one department to another contributing to an uninhibited exchange of bacteria and disease. The streets of Nizhneangarsk were full of stray dogs, garbage, and rats. The town's residents lived in wooden huts with no running water, plumbing, or sewer systems ... BAM workers lived in tents without heat or in barracks without running water or sewer systems. The melting of snow flooded the streets with fecal masses and even where there were outhouses, there was no sewage purification system. Because of these conditions, Nizhneangarsk became a pathogenic cesspool for staphylococcal disease and other infections.[14]

Shrayer and his colleagues turned to using phages from the Eliava Institute, then formally known as the Tbilisi Research Institute of Vaccines and Serums. The serum from the Institute, which was a cocktail consisting of numerous phages against different strains of staph, worked against 90% of the clinical isolates that doctors had collected. The BAM physicians used phages in cases of skin and wound infections, osteomyelitis and sepsis. "Polyvalent staphylococcal bacteriophage was applied with wads, bandages, sprinkles, and lotions during washing of lesions and abscesses (following pus removal), as well as intravenously, and with intratracheal administrations and intra-muscular injections," reports Shrayer. It was usually administered once a day for 5–7 days. And it seems to have worked, though Shrayer did not keep a control group. "Soon, patients' overall conditions improved, staphylococci stopped growing, wounds were without pus and began to granulate. The data of The Tbilisi Institute did not report any complications or side effects even among the doses administered intravenously," he writes.

The 1980s, just before the breakup of the Soviet Union, marked the heyday of phage use in the USSR. As Shrayer's experience shows, drug resistance was becoming a serious problem and, with the war in Afghanistan slowly draining the coun-

---

[13] Inside Russian Mecine, pg. 119.

[14] Shrayer, David. "Staphylococcal Disease in the Soviet Union: Epidemiology and Response to a National Epidemic,"Delphic Associates Inc. pg. 22 and 23.

try's economic resources, the Soviet government was unable to develop and manufacture enough new antibiotics to keep pace with the superbugs. Demand for phages, which are far cheaper than antibiotics to produce, grew. Staffers of the Eliava Institute began traveling widely across the USSR to cities like St. Petersburg, Saratov and Kazan, teaching physicians how to use their preparations. One of the Eliava Institute's biggest clients during that time was VAZ, an auto-manufacturing plant located on the Volga river in the city of Tolyatti, the Detroit of the Soviet Union. The plant had its own hospital with 3,000 beds, which had divisions of general surgery, gynecology, infectious diseases and ophthalmology, among others. "We went there two or three times a year and took samples from all parts of the clinic," says Zemphira Alavidze, a lead scientist at the Institute who used to head its oldest phage therapy lab. "We showed them everything, and they switched almost completely to phage therapy." Meanwhile, Teimuraz Chanishvili, who directed the institute until 2005, was putting the finishing touches on an IV phage treatment against staphylococcus that he'd spent 15 years developing. It would be a potent weapon against the emerging strains of staph that were resistant to two potent antibiotics: methicillin and Vancomycin. They had been purified, tested for toxins and shown to be effective in treating rabbits experimentally infected with staphylococcal sepsis. The preparation was also administered to 20 healthy human volunteers and shown to have no side-effects.

But before the treatment could be put into wide use, the USSR's highest legislative body, The Supreme Soviet, dissolved itself on December 26, 1991 and each of the former republics forged their paths to independence. Georgia's early years as a state were marked by violence and civil war, but the situation stabilized after Eduard Shevardnadze was elected president in 1992. Still, the Eliava Institute had lost its vast market. The center's manufacturing facility, which once employed some 700 people – the majority of the institute's staff – was shut down. Alavidze and some of the other microbiologists who ran phage therapy labs at the institute began doing their own manufacturing on a very small scale, producing an estimated 200–300 L per year, down from its 1980s peak of 200 L per day.

***

For over a decade, the Eliava Institute, now known formally as the Eliava Institute of Bacteriophage, Microbiology and Virology, was a shell of its former self. During my visit to Tbilisi in 2001, the staff fondly recalled the days when security guards waited at the front gate to check identification, and the lobby had attendants to check coats. The building's façade, supported by marble columns, used to be painted in bright colors, and sported a mural of Lenin, Stalin, Marx and Engels. In the lobby, the grand staircase leading up to the laboratories was covered with a rich red carpet, held in place by gleaming brass rods. But in 2001, the lobby was dark and cold for lack of heat or electricity – the only light came from the front door, which stood permanently ajar. The carpet was gone, and the marble floors and plaster walls were crumbling. Everywhere was the slightly sour smell of bacteria and agar.

Only the individual labs, of which there are ten, had a homier feel, though they all lacked heat and hot water. In Soviet days, the institute was renovated every year, but by 2001 it had been more than 10 years that much work had been done. Some of the scientists used their grant money to improve their labs and others did the work themselves: painting their offices, tearing down walls, cleaning their own bathrooms.

The scientists, too, were barely getting by. Like much of Georgia, which was destitute, they subsisted almost entirely on Western grants. In Soviet days, says Amiran Meipariani, a longtime colleague of Alavidze's, Institute scientists earned about 550 rubles per month, about four times the median salary in the USSR. "You could raise a family on that," he says. "You could go on vacation with your children, go to Moscow, travel abroad for 1 month." But in 2001, his base salary barely covered a month's bus fare.

The Institute's phage collection suffered along with the scientists. During the civil war that preceded Shevardnadze's election, fighting erupted in downtown Tbilisi, destroying buildings and wounding scores of civilians. Since there was sometimes no electricity at the Institute, the scientists took phages home with them and stored them in their own refrigerators. Despite heroic efforts, half the collection died.

In the United States, drug resistance is attributed to the use of antibiotics in meat and by the over-prescription of antibiotics by physicians. But in Georgia and Russia the overuse of antibiotics may be worse: many are available over the counter at low cost. A 2002 survey of Russian families found that 84% stocked leftover antibiotics in their home medicine cabinets, as compared to 26% of Americans.[15] Guram Gvasalia, a professor of surgery at the Tbilisi State Medical University and chief surgeon at one of the country's top military hospitals, says antibiotics like gentamicin, which Fred Bledsoe had received in the United States, worked on the vast majority of his patients in the 1970s and 80s, but today they work only 20% of the time. Newer drugs like ciprofloxacin, which was introduced in the 1980s, now work around 50% or 60% of the time, and the strongest antibiotics are too expensive for him and his patients to afford. Bacteriophages, which are available free of charge in the environment, are much cheaper.

Several recent Russian studies have reaffirmed the treatment's efficacy, especially in combination with antibiotics. Working in 2001 at Moscow's N. V. Sklifosovsky Research Institute of Emergency Medicine, part of the State Research Institute of Standardization and Control of Medical Immunobiological Preparations, Dr. E. B. Lazareva and a team of physicians tested bacteriophage on burn patients. In the study, 94 patients were divided into three groups. Of these, the biggest, constituting 45 subjects, received phages and antibiotics. The next group, of 40 patients, received only antibiotics, and the remaining nine patients received only phages. The phages, taken orally as pills, came from a Russian com-

[15] "The Inventory of Antibiotics in Russian Home Medicine Cabinets," L. S. Stratchounski et al. August 2003, electronically published by Clinical Infectious Diseases 2003; 37: 498–505.

pany called Biophag, located in the city of Ufa, and had been tested against clinical samples from the patients in advance. Patients took two pills, three times a day for 1 week, ingesting them 1–1½ h before mealtime. Each pill contained phages against five kinds of bacteria, including staph, strep and proteus.

The results were impressive, showing that antibiotics, when used in combination with phages, were more effective than when used alone. The overall condition of the phage and antibiotics patients improved and the microbial count of their wounds fell more precipitously than that of patients receiving antibiotics alone. Their wounds healed faster, and they were able to get their skin grafts earlier. Also, through blood tests, physicians found that phages were able to stimulate their subjects' immune systems more than antibiotics, by causing their bodies to release immunoglobulin A, an antibody that helps fight infections.

***

Although phage therapy has been portrayed in the press and in documentaries as a mainstream treatment in Georgia, in fact only around one in ten physicians still use it today. As an alternative therapy, it isn't taught at most medical schools, and many physicians have simply forgotten about it. Others believe it doesn't work, usually because they've misused it, say some of phage therapy's adherents. "If you decide simply to pour phages on a wound and dress it, nothing will come of it," says Gvasalia, the physician who treated Fred Bledsoe. Optimal results require physicians to work closely with microbiologists, sending bacterial samples to a lab to have the sample tested against and matched with a phage or group of phages. Then, a physician has to reapply phages at regular intervals, redress the wound and, if it doesn't heal, redo the testing to make sure a secondary infection hasn't taken root.

In their lab, Alavidze and her colleagues manufacture two types of phage preparations. One, called *seriuli-phagi*, or "serial phages," consists of mass-produced cocktails made up of viruses that target the most common gut or skin bacteria. The two major cocktails currently produced, called Piophage and Intestiphage – the names that d'Herelle originally assigned to them – patients can pick up at the small Diagnos 90 diagnostic lab on the grounds of the institute. The other types are customized phages, prepared for individual patients or for individual clinics and hospital wards. Because bacteriophages are so specific and because each strain of bacteria has so many serotypes (salmonella, for example, has more than 4,400 serotypes; Vibrio cholerae, which causes cholera, has 139), the customized phages are far more effective. Alavidze says that her "serial phages," are normally effective anywhere from 25 to 85% of the time. But, when tailor made for individuals or certain hospitals or wards, their effectiveness rises into the high nineties. That's because each health-care institution has its own strains of germs that grow, adapt and get passed around from environment, to health care worker, to patient.

The Eliava Institute has developed close relationships with several physicians who use phage therapy regularly, including Gvasalia, who has worked with Alavidze for more than 2 decades. The two have taken steps to develop new ways of administering phages to patients. In the late 1990s, Alavidze and Ramaz Katsarava, an expert on polymers at Tbilisi's Georgian Technical University developed PhagoBioDerm, a biodegradable dressing capable of gradually releasing phages into a wound. Gvasalia came up with the idea of making the bandage in bead form, so he could sprinkle it inside wounds and not have to reapply phages and redress the wound as often. He now uses this solid form of phages more frequently than the liquid form.

Gvasalia has been using phages on and off since 1975, alone, or in combination with antibiotics and immuno-stimulators – drugs that, similar to vaccines, lead the body to mount an immune response. "I always knew about phages," he says, "but I had my periods – like Picasso had his blue periods. Either I liked them or I forgot about them. And lately I've been using them. They're not a panacea, but they're a good treatment."

When I first met him, in 2001, Gvasalia had just treated a new mother for mastitis, an infection of the milk ducts. The woman had been on oral antibiotics for more than a month – antibiotics that she had bought at a pharmacy without a prescription – and, probably because the antibiotics were the wrong type for her infection, developed an abscess that began leaking pus. Gvasalia drained her infection and sprinkled PhagoBioDerm inside. Ten days later, the infection was gone and the wound had healed.

Across town from the Republic Hospital is the small Genesis clinic, where Chief Surgeon Ruben Kazaryan was treating adolescents and children who had been displaced by the secessionist conflicts in the Georgian republics of Ossetia and Abkhazia. The clinic is clean and modern, with pale aqua tiles on its floors and a fluorescent-lit lobby. Kazaryan is small and energetic, with dark shaggy hair. In rapid-fire Russian, he described his mistrust of antibiotics. "My opinion of antibiotics was always cautious," he says. "I use them only in life-or-death cases where there is no other choice. When you explode an antibiotic in your system, it kills everything around it, both bad and good. Antibiotics are immuno-depressing, because they work instead of your body." Kazaryan's belief is that phages help a person's natural immune system to conquer an infection on its own. "Phages decrease the number of bacteria to the point where the immune system can finish the job by itself," he says. He was using phages for infections of the ear, sinus and throat, for burns and for any chronic bacterial infection. For tonsillitis, he would rinse the tonsils with phages. One of his most recent patients had a sinus infection. He ran 3 days of tests and found that she had a mixed infection of staph and strep that was sensitive to a pre-made Eliava cocktail called Enkophag. He sent her home with a 10-day supply of phage ampoules and told her to drop 1 mL into her nose twice a day. She'd have to hold her head back for 1 min and breathe through her mouth, and then spray additional phages into her throat, gargle and swallow. "Antibiotics are still the standard treatment in Georgia," he says. "But my determination and conviction is that the era of antibiotics is over. They must exist, but not everywhere. Why are people cautious about using phages, when they should be cautious about using antibiotics?"

***

Today, the Eliava Institute's fortunes have improved. Since 2006 it has been directed by Rezo Adamia, who headed the institute's molecular biology lab before being appointed a Member of Parliament in 1992 and then Georgia's ambassador to the U.N. in 2002. Under his leadership, the institute became independent, moving out from under the Georgian Academy of Sciences, and began spinning off companies to generate revenue. He reports that the Institute's annual income has risen from just $50,000 in 2006 to about $300,000 today. Phage production is up, and the Institute recently opened an international outpatient clinic on its grounds.

Ongoing grants from several sources, including the U.S. State Department, the U.S. Department of Defense, NATO and the International Science & Technology Center, which supports scientists in the former Soviet Union, allowed the main building to be renovated in time for an international phage conference in 2008. Georgian President Mikheil Saakashvili, whose grandmother worked as a microbiologist at the Eliava Institute for 30 years, attended and spoke about her experiences.

In 2005, Gvasalia, with financial support from Kutter's PhageBiotics Foundation, started a training program for newly minted M.D.s, teaching them how to use phages to combat such conditions as skin infections, urinary tract infections and prostatitis. A recent graduate of that program, Lasha Gogokhia, is now completing a postdoctoral fellowship at the University of Utah in Salt Lake City and plans to apply for a surgical or infectious disease residency in the U.S. A practice that once seemed on the verge of dying out is being passed down to a younger, more cosmopolitan, generation.

Chapter 6
In Poland: Phages for Diabetes?

Kazaryan's faith in phage therapy is shared by a small but equally passionate group of physicians and researchers in Poland. Wroclaw, a Silesian city of gothic churches and cobblestone streets, is home to the L. Hirszfeld Institute for Immunology and Experimental Therapy of the Polish Academy of Sciences, the only institute outside the former Soviet Union that regularly provides therapeutic phages for patients. Housed in a 1970s-style concrete-and-glass building in the green outskirts of Wroclaw, the Hirszfeld Institute is known throughout Western Europe for its research on stem cells, bone marrow transplants, and on the development and selection of T cells, a type of white blood cell. Phage therapy makes up a small but growing segment of its work.

Over the last two decades, the Polish center has collected and published an impressive volume of data on the treatment, in the Institute's own English-language journal, Archivum Immunologieae et Therapiae Experimentalis, as well as in other Western journals. The studies described in the papers aren't immune to criticism, even self-criticism: the Hirszfeld Institute has almost always done its research studies in the absence of double-blind controls – a fact its scientists regret and attribute to their lack of resources. But the sheer quantity of cases presented, combined with the fact that virtually all the cases involve patients who failed to respond to antibiotics, is persuasive. The Institute's self-reported success rate as of 2000 was 86% in 1,300 documented cases.[1]

The most striking of the Institute's studies is one on antibiotic-resistant septicemia, published in 2003 in the American journal Transplantation Proceedings. Septicemia is a bacterial infection of the blood that disproportionately affects those with weakened immune systems, such as newborn babies, the elderly, cancer patients and those with diabetes. It affects some 500,000 people in the United States annually and has a mortality rate of 30–50%. Increasingly, this infection is caused by germs that have grown resistant to antibiotics – in many cases, patients contract septicemia in the hospital where drug-resistant germs can be especially prevalent.

---

[1] Weber-Dabrowska, B. et al. (2000) "Bacteriophage therapy of bacterial infections: an update of our institute's experience," Archivum immunologiae et therapiae experimentalis (Warsz) 48: 547–51.

A. Kuchment, *The Forgotten Cure: The Past and Future of Phage Therapy*, DOI 10.1007/978-1-4614-0251-0_6, © Springer Science+Business Media, LLC 2012

In the study, 94 septicemia patients who had not responded to antibiotics were given phages orally for an average period of 29 days. The largest number of patients (36) had contracted the infection as a complication following surgery; others had developed it as the result of injury, urinary tract infection, pneumonia and peritonitis. Following treatment with phages, 85% recovered fully; the rest showed no response. Clouding these results is the fact that, because many physicians were reluctant to take their patients completely off antibiotics, 71 patients received phages and antibiotics together and 23 received phages alone. But, the paper reports, the group that received both treatments had the same rate of recovery as the one that received phages alone, suggesting it was the phages and not the antibiotics that cured them.

More recently, the Institute has published studies on chronic bacterial prostatitis, a hard to treat infection of the prostate gland. Because many antibiotics have trouble penetrating deeply enough into the tissue, men can struggle with this condition for months. In 2009, the Institute reported a 50% cure rate in a group of 22 men with antibiotic-resistant infections.[2] Gorski is currently overseeing the start of a placebo-controlled clinical trial for this ailment to further test phage therapy's effectiveness against it.[3]

The Hirszfeld Institute is a modern, well-staffed facility, with a total of 300 workers. The phage therapy lab, whose stock of phages is managed by a tall blond woman named Beata Weber-Dabrowska, operated in much the same way as Alavidze's lab in Tbilisi when I came to visit in 2003. Whenever a patient or local doctor sent in a bacterial sample, Dr. Weber Dabrowska and her staff would head to the institute's ground floor, where a door with a bar and a padlock across the front guards her phage bank. Inside the walk-in refrigerator are some 300 individual types of phages for bacteria like staph and *E. coli*. They're kept in glass flasks or in small glass ampoules, sealed by hand with a blowtorch by one of Weber-Dabrowska's workers upstairs, and stacked in small, white cardboard boxes. Each time she receives a sample of a particular bacteria, Weber-Dabrowska tests all the phages in her arsenal that are specific to that particular strain. For *E. coli*, for instance, she has some 80 different phages. Her staff would meticulously screen each one against the bacteria. They do this by placing the bacterial sample in a test tube filled partially with broth. They then place the test tube in a rack and place the rack inside an oven set to 37°C (98.6°F) in order to allow the bacteria to propagate. After 4 or 5 h, the bacteria will have multiplied inside the test tube, turning the formerly clear golden broth cloudy. The contents of the test tube are then spilled over the surface of an agar-coated Petrie dish and allowed to dry for an hour or two in the incubator. Then, a lab worker takes a magic marker and divides the lid of the Petrie dish into six sections resembling slices of pie. She labels each 'slice' according to the type of phage

---

[2] Poster session, "Pathogen eradication by phage therapy in patients with chronic bacterial prostatitis." Slawomir Letkiewicz et al., Poster number 374, the 25th Annual European Association of Urology Congress, Barcelona, 16-20 April, 2010.

[3] Interview with Andrzej Gorski, May 26, 2010.

she plans to drip onto each section. The phages kept in the lab are stored in tiny glass bottles capped with an eyedropper. The lab worker places one drop of each phage onto the Petrie dish and places it back into the incubator overnight. The next day, they read the results. When a phage is active against a particular bacterial strain, it will clear a circle, sometimes as large as a quarter, in the cloudy bacterial film that covers the surface of the dish.

For now, the Hirszfeld institute produces only liquid phages that can be applied topically or taken orally. Andrzej Gorski, who oversees the institute's phage work, and Weber-Dabrowska have not yet found the resources to produce a preparation pure enough to be injected via IV, although this is one of their future goals. Says Dr. Weber-Dabrowska, "Can you believe that in the 1930s phages were administered intravenously, intraperitoneally and even into the heart? [one can] imagine that this will be possible once again."

***

The Hirszfeld Institute's fate is closely intertwined with the history of Jews in Wroclaw, and it may never have come into existence were it not for Stalin's sudden death in 1953.

Until the end of World War II, Wroclaw was part of Germany and had a solid German population, with a large Jewish minority. The city had been Polish in the Middle Ages, but had then become part of Bohemia, the Habsburg Empire and Prussia. When the Red Army captured it from the Nazis in May, 1945, more than half the city was destroyed in the siege but has since been meticulously restored. Only a few charred and pockmarked walls stand as memorials.

More than 90% of Wroclaw's Jewish population was killed during the war. But a few managed to survive in hiding or by living on forged documents. In 1945, the Soviet Union annexed an Eastern slice of the country, including the now Ukrainian cities of Lvov and Vilna, and expelled some 5 million Poles into modern-day Poland. At the same time, the USSR pushed Poland's boundaries westward, incorporating the region of Silesia, of which Wroclaw is the capital, into Poland and exiling the Germans living in those areas to the GDR. Wroclaw, cleansed of Germans, welcomed instead a flood of Poles from the pre-war east, including much of the medical faculty of the University of Lvov, who took up positions at Wroclaw's renowned Academy of Medicine.

Wroclaw, famous for the 100 bridges that straddle its Odra river, has an impressive academic tradition. Its Academy of Medicine, founded in 1811 at what was then Breslau University, became independent in 1950. Many famous physicians, including the neurologists Alois Alzheimer and Karl Wernicke worked and taught there. Today, one seventh of Wroclaw's current 700,000 inhabitants are students who help attract foreign companies like Bosch, ING and Toyota. The investments have helped turn Wroclaw into one of Poland's most vibrant post-Communist cities, with one of the country's lowest unemployment rates.

The story of the L. Hirszfeld Institute is also intimately tied to the personal history of Ludwik Hirszfeld, who was to become one of the most prominent immunologists in Europe. Born in 1884 in Warsaw, which was then occupied by Russia, Hirszfeld graduated from the University of Berlin in 1907, then did important work on the heredity of blood groups with Emil von Dungern, one of the creators of the science of immunology and a professor at Heidelberg's Institute of Cancer Research. With the outbreak of the war in 1914, he and his wife, Hanna, who was also a physician, volunteered to serve in the Serbian Army, helping to battle various epidemics, including typhoid, that were running rampant in the Balkans. Upon returning to Poland, both Hirszfelds converted from Judaism to Catholicism to avoid anti-Semitism and to prove their devotion to Poland. But that did not save them from tragedy. With the outbreak of World War II, the Hirszfelds, along with their then 20-year-old daughter Marysia, were rounded up along with other Jews and forced to live in the Warsaw Ghetto. There, he worked to control the spread of typhoid and taught in the underground medical school. Still, many Jews resented him for his conversion. In 1942, when the mass deportations of Jews from the ghetto to concentration camps started, the Hirszfelds escaped. With the help of non-Jewish friends, the family hid – sometimes separately, sometimes together – and existed on forged documents. But they did not escape the Ghetto unscathed: the Hirszfelds' daughter, Marysia, had fallen ill in the ghetto, suffering from anorexia nervosa, a condition in which the patient is too traumatized to eat. She died in 1943, hiding at an estate that belonged to Polish gentry. Wrote Hirszfeld in his 1947 memoirs "A History of One Life": "All the work of my life ... just sufficed [to permit] my child to die in her 23rd year of life in a bed ... to be buried in her own grave, even though under a false name. Under her own name she had no right to live or die."

After the war, the Hirszfelds settled in Wroclaw, where Ludwik accepted positions as dean of the medical school, which was then still a part of the University of Wroclaw, and chairman of the school's department of microbiology. In Wroclaw, Hirszfeld initiated mass testing and treatment of syphilis, which was rampant at the time, due to the mass population shifts and troop movements following the war. Some 200,000 subjects were tested and around 1,000 patients were diagnosed and treated. Hirszfeld also continued work he had begun earlier on blood groups, paying special attention to the blood types of pregnant women and fetuses. He postulated correctly that frequent miscarriages could be the result of conflicting blood types between mother and baby.[4]

In 1952, Hirszfeld petitioned the Polish Academy of Sciences to set up an institute devoted to immunology within the Medical School's Department of Microbiology. The request was possibly slowed by a series of events taking place in the Soviet Union at the time. One was the triumph of the misguided scientific philosophy of Lysenkoism, which Stalin staunchly backed. (Incidentally, d'Herelle had also been a firm believer.) Trofim Lysenko was a Russian biologist and agronomist who maintained, like Lamarck, that acquired traits are inherited. He also denied the

---

[4] My Association with Ludwik Hirszfeld, Wroclaw 1945–1954, by Felix Milgrom, Archivum Immunologiae et Therapiae Experimentalis, 1998, 46, 201–212.

existence of the gene, which he termed "alien foreign bourgeois biology."[5] Lysenkoism grew widespread, even in Soviet satellite countries like Poland, because any biologist who failed to publicly embrace the philosophy risked losing his job. This "Lysenkomania" damaged Hirszfeld, since his major discovery had dealt with the genetic inheritance of certain blood groups. Writes Felix Milgrom, a close associate of Hirszfeld's, "Once Hirszfeld told me that he would be ready for any possible martyrdom if by sacrificing himself, he could help in freeing the Soviet Union and the People's Republics from bondage to Lysenko's biology."[6]

The same year that Hirszfeld petitioned the Polish Academy of Sciences, a dozen prominent doctors in Moscow, most of them Jewish, were accused of attempting to poison high army and party officials. In what came to be known as the "Doctors' Plot," all of the accused confessed to their crimes, renewing vicious persecution of Jews in Soviet science and medicine. "We anticipated that similar persecution and accusations would follow in other People's Republics including Poland," writes Milgrom. "Undoubtedly we were first in line to be targeted, since many senior staff members in Hirszfeld's department were of Jewish origin. Furthermore, we had ample contacts with the West and published many papers abroad. At that time we all felt quite uneasy, or to put it bluntly, scared." Then, suddenly, Stalin died in early 1953, the doctors were freed, and the political situation began to gradually change. Whether for this reason or others, Hirszfeld's petition was granted in 1954, and the Institute of Immunology and Experimental Therapy was officially created.

Unfortunately Hirszfeld did not live long enough to direct his new institute. He died in March of 1954, of complications following surgery. Though he had designated Milgrom, who was also Jewish, as his successor, the Polish Academy of Sciences handpicked another: a young Polish immunologist named Stefan Slopek, who lead the Institute into the 1980s and became Poland's "father of phage therapy," says Weber-Dabrowska. Milgrom emigrated to the United States, where he lived in Buffalo, New York and was professor emeritus at the State University of New York, in the department of microbiology, until his death in 2007.

Throughout the 1950s and 1960s the Institute was involved in phage typing: using phages to identify bacteria by seeing if a known phage would lyse them. "We had obtained more than 2,000 or 3,000 Shigella strains," recalls Weber-Dabrowska, who has worked at the institute for more than 30 years. "And we had an international set of phages for these strains." The Institute of Immunology had traded phages with research centers in nearby countries, including Sweden, Russia, Bulgaria and Hungary, as each one was working on the pure science of typing. Back then, phages drew little interest as a therapy, and held no proprietary or commercial value – they were used primarily for diagnostic and research purposes.

---

[5] "Lysenko, Trofim Denisovich *A Dictionary of Scientists*. Oxford University Press, 1999. *Oxford Reference Online,* Oxford University Press, 15 July, 2003.

[6] Milgrom, p. 208.

That changed in the early 1970s, when an epidemic of Shigellosis, or dysentery, swept local orphanages. Dysentery, caused by one of several types of Shigella bacteria, causes severe abdominal cramps, bloody diarrhea and high fevers; young children, especially those already suffering malnutrition, have the highest mortality rates from the disease. The Institute used its vast stock of Shigella phages to treat patients and help stem the disease's progress.

Building on its work with Shigella, the institute branched out into fighting antibiotic resistant infections, which were just beginning to emerge. Under Slopek's rule, phages were provided free of charge to patients, the money coming out of the lab's own funding from the Polish Academy of Science (PAN) and from grants. But, starting in the 1980s, patients had to pay for their therapy out of their own pockets, after the government mandated that Ministry of Science funds could go only toward research and not treatment. And, unlike antibiotics, phages are not covered by Poland's national health insurance. "For 10 years," says Weber-Dabrowska, "phage therapy was conducted on a very small scale and that's why it wasn't able to move forward very rapidly. We had only to survive, and I did my best to keep the phages in good order and help people."

In 2005, after Poland joined the European Union, the Institute received formal permission to open a 5-room outpatient clinic to treat patients with antibiotic resistant infections. In accordance with international guidelines known as GMP, or Good Manufacturing Practice, the Institute contracted with a local vaccine company to produce phages under strictly controlled conditions. Still, the drug is categorized as an experimental treatment and it will remain that way until the Institute, or another group of researchers, complete double-blind clinical trials.

Across Poland only around 100 doctors regularly use phages, Weber-Dabrowska estimates. Still, like in Georgia, the doctors who use phages are a devoted and passionate group. Maciej Dworski, a small, genial general surgeon with a mustache and a ruddy complexion, has been using them for the last 20 years, ever since he managed to cure a particularly sick patient: a middle-aged woman with septicemia, a bacterial blood infection, which she caught from a contaminated hospital IV. The infection was so bad that, despite treatment with antibiotics, she developed abscesses in her lungs and near her eyes. Desperate, the head of Dworski's department traveled 50 km to Wroclaw from Jelenia Gora, a picturesque town at the foot of the Karkonosze mountains, for help. The institute provided phages for the woman to take orally, and she recovered within a few weeks.

Dr. Dworski, who spent most of his career at Jelenia Gora's County Hospital but now lives in Wroclaw and works for the Polish National Health Service, has used phages as more than a last resort. Over the years, he identified two common conditions for which he would automatically administer phages: breast infections in new mothers and foot infections. In the United States, up to one-third of new mothers develop breast infections, which can appear suddenly and, if not treated quickly, can require surgical drainage and an overnight hospital stay.[7] The condition occurs when

---

[7] "Lactation Mastitis" Cibele Barbosa-Cesnik et al., JAMA 2003;289:1609–1612.

bacteria from the baby's mouth enters the milk duct through a crack in the mother's nipple. Dr. Dworski says that, thanks to phage therapy, he no longer needs to make incisions into even the most complex cases of mastitis. His procedure is to drain the site of the infection by pricking it with a needle and then to inject it with phages. Women are treated for a period of 2 days to 2 weeks, coming into the office for daily injections and may continue breast feeding. "The results," he says, "have been fantastic." In cases of minor foot infections – most often caused by ingrown toenails, the procedure is similar: drainage, followed by local application of phages. Still, most Western doctors will tell you that simply draining puss from an infection is often enough to cure it – even without phages or antibiotics.

Andrzej Gorski arrived at the Institute as director in 1998 and has made expanding the institute's work in phage therapy a priority. He has since stepped down as director to focus on overseeing phage work. Gorski, a member of Poland's Academy of Science, is an experienced internal medicine specialist with a subspecialty in clinical immunology. He trained at New York City's Sloan-Kettering Institute for Cancer Research and Seattle's Fred Hutchinson Cancer Research Center. Before coming to Wroclaw, he was president of the Medical University of Warsaw but left after his first term when he learned of the opening at the Hirszfeld Institute. His decision to come to Wroclaw was personal – connected with his own family's complicated history and his admiration of Hirszfeld. "I felt it was a moral obligation," he says.

At the time, Gorski was only dimly aware of phage therapy, but it quickly captured his interest. "I was a practicing physician in Warsaw for 20 years and took care of patients after kidney transplants, so I saw the problem of antibiotic resistance" he says. "I realized that the problem of antibiotic resistance will be increasing in the future, and I saw a great potential for using phages for that purpose. It was a decision based on my instinct and experience." He built a new phage therapy lab for the institute and became its titular director, so as to give it his MD qualifications. He has personally reviewed and helped edit all their publications. It's part of his broader effort to put more emphasis behind the Hirszfled Institute's applied sciences, in order to bring more money to the institute and to help it expand.

Although he is eager to win more publicity for the institute, he is also cautious about the type of media coverage he attracts. "We have to be very careful," he told me on more than one occasion. Some of the hesitation stems from the fact that he is new to phage therapy and that he feels responsible for a field that took his predecessors some 30 years to establish. He does not want to, with one false move, shatter its prospects. "As a successor to L. Hirszfeld, I have a duty to further promote his work and achievements, and be sure that full credit is given to the accomplishments of the Institute which he founded just before his death," he wrote in an e-mail. He has been displeased with much of the coverage phage therapy has received in the United States, where journalists, he feels, have overhyped it and have failed to mention his institute by name, attributing its achievements vaguely to "Eastern Europe" or "doctors in Poland."

During my 2003 trip, we paid visits to the deputy mayor of Wroclaw and to the chancellor of Lower Silesia to boost awareness of the institute in general and of phage therapy in particular. One of Gorski's dreams is to open a phage therapy institute dedicated to the study of bacteriophages and their clinical applications. In recent years, the Phage Therapy Lab has been exploring how phages interact with a variety of biological systems, not just bacteria. For instance, Gorski's team has found that phage therapy exerts a strong anti-inflammatory action, reducing levels of so-called C-reactive protein or CRP. The inflammation process, marked by high levels of CRP, among other signs, triggers many of the symptoms associated with bacterial infections: pain, swelling, mucous production, breathing problems. "CRP may drop during phage therapy even in patients in whom eradication of the infection has not been achieved," says Gorski.

He is also intrigued by recent studies in the journals *Nature* and *Science* showing a link between certain gut bacteria and conditions like obesity and diabetes. "This opens up a very exciting field of research," says Gorski. "Would you ever have thought to treat, for instance, diabetes with phages?" Perhaps his institute will one day prove that's possible.

Chapter 7
The Renaissance of Phage Therapy

One morning in 1993, Carl Merril read an alarming article in The Washington Post. It told of the growing problem of antibiotic resistance, a story that had been creeping into the press since the early 1980s. From Tokyo to New York City, the Post wrote, hospitals were contending with a new breed of superbugs, bacteria that had grown immune to the strongest of medicines – a situation that had resulted from the overuse of antibiotics. Though Merril was well aware of the problem, this lengthy report was particularly disturbing. "We're starting to see organisms that we can't treat with anything," John G. Bartlett, chief of infectious diseases at Johns Hopkins Hospital in Baltimore, told the newspaper. "What do we do? We give drugs and we pray."[1]

Merril, a longtime biochemist at Bethesda's National Institutes of Health, put down the paper and went to take a hot shower. "Showers are the best place for me to get ideas," he says. There, Merril began thinking back to a class he took in 1965 after finishing medical school: the legendary Phage Course taught each summer on the picturesque grounds of the Cold Spring Harbor Lab. This memory would eventually lead Merril, an affable, energetic presence, with reddish hair and oval spectacles, to embark on a set of experiments that would help propel phage therapy back into the pages of Western medical literature after an absence of 50 years.

In the 1960s, Merril was working as a research associate at the National Institutes of Mental Health, the NIH division from which he has since retired, and a colleague told him about the course. By the time Merril arrived at Cold Spring Harbor, Max Delbruck had long since moved on; Al Hershey was the only one who remained, though he rarely made an appearance. And the mood seemed to have sobered up. "It was very intense," Merril recalls. "You did a lot of experiments late into the night and got up early in the morning to see what the results were before attending lectures and lab demonstrations."

---

[1] "Running Out of Wonder Drugs," by Sandra G. Boodman, The Washington Post, March 16, 1993, pg. Z10.

A. Kuchment, *The Forgotten Cure: The Past and Future of Phage Therapy*,
DOI 10.1007/978-1-4614-0251-0_7, © Springer Science+Business Media, LLC 2012

Although the Phage Course did not concern itself with the therapeutic use of phages, Merril couldn't help but wonder about the possibilities. He decided to pose the question to one of his instructors.

"If these viruses can kill bacteria, why aren't we using them to treat infectious diseases?" he asked.

"Go read Arrowsmith," the instructor replied.

Others referred Merril to Gunther Stent's "Molecular Biology of Bacterial Viruses."

Not satisfied with these responses, Merril decided to see for himself if, as Stent hypothesized in 1963, gastric juices and antibodies would prove to be the enemy of phages. When he returned to Bethesda from Cold Spring Harbor, Merril and his colleagues took a colony of mice and injected them with phages to see how long the viruses would linger in their system. As Merril had expected, these particular phages began to vanish from the animals almost as soon as they were introduced – but not because of antibodies or gastric juices. Before the mice even had time to develop antibodies, their livers and spleens – known as the reticuloendothelial system – had snared nearly all of them. Because the bugs were new to the animals' bloodstream, they had not yet learned to evade these organs. "Was this why phage therapy failed?" Merril wondered.[2] He and his colleagues published the study in Nature in 1973 and then moved on to other subjects. In the United States at least, antibiotics were still powerful wonder drugs and there was no need to come up with alternatives.

Standing in the shower 20 years later, Merril thought back to his early work. And suddenly an idea occurred to him: perhaps there was a way to breed more-effective phages that could outwit the liver and spleen. He began sharing his brainstorm with coworkers at the NIH, all of whom encouraged Merril to try it out. At the same time Richard Carlton, a New York psychiatrist and founder of the technology transfer company Clinical Trials Consulting, was looking for a new project in which to invest. He had just helped a group of researchers sell a new drug for septic shock to a pharmaceutical company and was on the lookout for a promising new technology that he could help take from research to development. A common acquaintance put them in touch.

At first, Carlton wasn't sure what to make of Merril's idea. On the one hand, it sounded promising, and Carlton, like Merril, was greatly troubled by the problem of multidrug resistance. On the other hand, phages were unproven and tainted by their long association with Soviet medicine.

What sealed Carlton's decision to work with Merril was a passage he'd read in one of Felix d'Herelle's books, "The Bacteriophage and Its Clinical Applications." Initially, Carlton had been put off by d'Herelle's boastful tone in describing the unqualified success of phage in curing diseases like cholera, dysentery and bubonic plague. But then he came upon a disease that had confounded d'Herelle: typhoid fever. "Contrary to what might be anticipated in view of the effectiveness

---

[2] Mark R. Geier, Michael E. Trigg, Carl R. Merril, "Fate of Bacteriophage Lambda in Non-Immune Germ-free Mice," Nature Nov. 23, 1973 (246). Pg. 221–223.

of bacteriophage therapy in other infections of the intestinal tract, such as cholera and dysentery, the use of bacteriophage in the typhoid and paratyphoid infections of man has not been attended by any considerable success," d'Herelle wrote. "... As a general rule most investigators ... have failed to find any evidence of therapeutic efficiency, and during the past year, in connection with an outbreak of typhoid fever in Lyons, studies made under my direction have given entirely negative results ... This but serves to emphasize the fact that our knowledge with regard to infections of this type is far from complete."[3] "It had the ring of truth," says Carlton. In fact, experiments performed since d'Herelle's time have shown that specific phages can be highly effective against typhoid.[4] The reasons behind their efficacy, however, remain something of a mystery. The bacteria that cause typhoid, *Salmonella typhi*, burrow inside immune cells called macrophages – cells long thought to be beyond the reach of phages. Scientists are still trying to puzzle out whether phages cure typhoid by attacking bacteria inside macrophages or by attacking only the bacteria *outside* the macrophages. If it's the latter, then phages would work by lowering the overall bacterial count in the body and helping the patient fight off the infection on his or her own.[5]

Carlton also felt heartened after reading some of the studies out of the former Soviet Union and Poland; though they weren't double-blind studies, they were promising. "So, I held my nose and jumped," he says.

Merril and Carlton began assembling a team to carry out the initial lab work. Merril enlisted his friend Sankar Adhya, a noted molecular biologist at the National Cancer Institute and a member of the National Academy of Sciences, to help him design the experiment, and the three put together a proposal for a CRADA – a cooperative research and development agreement that allows scientists in the public and private sectors to work together. The proposal was approved based on scientific merit and a lack of conflicts of interest, and the partners headed for their labs.

The team knew there would be many hurdles to overcome. First, there was little recent evidence that phages could rescue animals. The only reliable tests performed since the 1940s on infected animals were conducted by two British researchers in the early and mid 1980s. H. W. Smith and M. B. Huggins of the Houghton Poultry Research Station in Houghton, England successfully treated mice and calves with experimental *E. coli* infections using phages – and even claimed to have obtained better results with phages than with common antibiotics like tetracycline and ampicillin. Moreover, they found that bacterial resistance to phages was not a significant problem; sensitive bacteria continued to outnumber resistant bacteria, and of the germs that had developed resistance to phages, many had mutated into a harmless,

---

[3] D'Herelle, Felix. "The Bacteriophage and Its Clinical Applications." Trans. George H. Smith, (Springfield, Illinois: Charles C Thomas, 1930. pg. 185–186.

[4] See Ward, W. E. (1943) 'Protective Action of Vi bacteriophage in Eberthella typhi infections in mice,' Journal of Infectious Diseases, 72: 172–6, cited in Hausler, Thomas "Viruses Vs. Superbugs" Macmillan 2008, pg 114.

[5] Hausler, Thomas. Trans. Karen Leabe. "Viruses vs. Superbugs: A Solution to the Antibiotics Crisis?" (NY: Macillan, 2008), pg. 114.

or avirulent, form. But, says Merril, the attitude toward phage therapy in the United States was so negative that many scientists refused to believe the results. Merril, too, came under sharp criticism from his most hardened colleagues. "Aren't you ashamed to do this kind of work?" one associate asked him point blank. "You of all people should know that the bacteria are going to become resistant, and it's going to be worthless." Merril told me later via e-mail, "This argument is correct – in the long term- just as it is for all of our antibiotics. However, until the resistance develops one can save a number of human lives. In addition, with our knowledge, we can rationally continue to develop phage strains to overcome such resistance, just as the chemists are endeavoring to do for the antibiotics." Merril, Carlton and Adhya were undeterred by the negativity. They hoped they'd find that bacteriophage therapy had been unjustly marginalized in the West. Perhaps they'd be the ones to restore it to its proper place in medicine.

Within a few months, their team in place, the experiment was ready to go forward. First, they worked on breeding a phage that could evade the liver and spleen. Merril and his team of researchers injected phage active against a specific strain of *E. coli* – a bug best known for contaminating undercooked meat and causing severe food poisoning – into the stomachs of mice. They chose this specific type of phage, called *lambda*, because it was among those that Delbruck's group had studied the most thoroughly. After 7 hours, Merril and his team took blood samples from the animals, isolated the phages still present, propagated them and reinjected them into the mice. After performing this task, called serial passage, eight times, the team managed to isolate several phages that could remain in the bloodstream for as long as 18 hours. The survival rate of these new phages was thousands of times higher than that of the "parent," lambda. In other words, if you injected 100,000 of the newly bred phages into a mouse, 62,500 would still be alive 18 hours later, as compared to just a single wild-type phage. They called these new phages Argo1 and Argo2, after the ship sailed by Jason and the Argonauts in Greek myths. "The ship navigated all those treacherous passages and still kept floating," said Carlton, "And these phages just kept sailing past the spleen and the liver."

Next, they would have to see if Argo1 and Argo2 could actually save mice infected with *E. coli*. They took four groups of mice and injected each one with a lethal dose of bacteria. Thirty minutes later, three of the groups were treated with phage and the last was left untreated as a control. The results were dramatic: all of the untreated mice perished in 2 days. Among those that were treated with phage, all survived, and the ones injected with Argo1 and Argo2 had a milder illness before recuperating than those treated with the wild-type (non passaged) strain of lambda. The results were as good as anything the trio had hoped for. "We weren't sure we could get a phage to stay in the circulation longer and we didn't know whether phage could really rescue animals," says Merril. "And, lo and behold, it did both."

Merril, Carlton and Adhya weren't the only ones excited about their results. When they submitted their paper to the Proceedings of the National Academy of Science, the editor wanted to draw special attention to their findings, so he commissioned the Nobel laureate Joshua Lederberg, one of the most respected names in bacteriology and public health – and who had made repeated warnings about the dangers of antibiotic resistance – to write an accompanying commentary. He met with Merril, Carlton and Adhya, pored over Merril's past phage work, and highly praised their experiment, calling it "an ingenious surmounting of one of the hurdles to the use of phage in therapy." He went on to remind readers, as Stent had, of potential obstacles to the treatment, including an immune reaction and the problem of bacterial resistance, but suggested how some of these problems might be overcome: bacterial resistance could be delayed with the use of a phage cocktail – pitting several different types of phages against a single bacterium – and, "in these days of genetic engineering, other tricks come to mind." He also noted some specific uses for phage in medicine: "If the transmission of cholera or dysentery could but be mitigated, the refugee camps bordering Rwanda might be less terrifying. Yes, d'Herelle had enunciated these dreams, and today we are somewhat put off by the anecdotal quality of the evidence that he assembled. Even so, it is ambitious but not preposterous to suggest that newer knowledge might yet engender some workable weapons for the medical armamentarium." Merril and his partners were thrilled with the review. "If my mother had read our PNAS paper, she couldn't have written a better review than he did," Merril joked.

The two papers generated a flurry of interest in phage therapy. Merril, Adhya and Carlton patented their serial passage technique, and Carlton invested $50,000 of his own money to build a company around the technology. The team's next step would be to test bacteriophages against a microbe far more deadly than *E. coli*. They turned their attention to VRE, or Vancomycin-Resistant Enterococci, one of the first drug-resistant bacterial strains to appear in hospitals in the late 1980s. The bacteria *Enterococcus faecium* is part of the normal intestinal flora of humans and animals, and it doesn't cause problems unless it happens to leak out into the blood stream. When this happens in healthy patients, white blood cells descend on the bugs like piranhas and destroy them. But, in patients with weak immune systems – such as transplant recipients, cancer patients undergoing chemotherapy or those with diabetes – the bugs can multiply in the bloodstream and invade vital organs. When an Enterococcus infection turns into bacterimia – a bacterial blood infection – it has a fatality rate of about 37%. Enterococci can also cause hospital-acquired, or nosocomial, infections. "This bug is very hearty in the environment," says Carlton. "It survives on surfaces such as stethoscopes, EKG knobs, curtains. It's all over the place." In the hospital environment, enterococci encounter Vancomycin, the most common antibiotic deployed against it, and mutate into a resistant form. In this monster version, VRE can cause urinary-tract infections if it contaminates a catheter or endocarditis if it contaminates equipment used in angiograms, or it can colonize patients' guts and then seep out to cause bacteremia. Carlton, Merril and Adhya set out to see how phages would fare against this germ.

In a new experiment, they infected five groups of mice with VRE they had isolated from the blood of an infected hospital patient. To tackle the VRE, they had isolated several corresponding phages from a sewage treatment facility near Bethesda. Nearly each week, the researchers would visit the Seneca treatment plant, don disposable paper clothing and venture down to a pipe that spat out raw sewage, and collect it in plastic containers for analysis. The stomach-turning work paid off: they managed to find phages that killed a high percentage of clinical strains of VRE in the lab; and, this time, they didn't have to do serial passage to make the viruses work well in mice.

After injecting VRE into the abdominal cavities of five groups of mice, the researchers administered phage to all but one group. In the untreated group, all the mice died within 48 hours. But in the two groups of mice that received the highest doses of bacteriophages, all survived and experienced only a mild illness. The researchers found that bacteriophage could rescue mice when injected from 45 min to 5 h after they received VRE injections, and that even when phage was administered to moribund mice, it managed to rescue half of them.

The researchers took extra care to address previous findings about phage therapy. For instance, they proved that the mice recovered because of the bacteriophages' ability to destroy the VRE, and not because of an indirect immune response, as scientists in the 1930s and 1940s had suspected. They demonstrated this fact by administering heat-killed phages to the mice – a heat-killed phage would generate an immune response but would not lyse the bacterial cells. These mice perished at the same rate as those that had received no phages at all.

Carlton named his company, which he based in Port Washington, N.Y., Exponential Biotherapies. In 2001, he completed Phase I clinical trials on his VRE preparation in England, under the auspices of the International Committee on Harmonization. Thirty healthy volunteers received injections of VRE phage twice a day for 9 days at various dosages. There was only one side-effect: a mild rash in a single volunteer after the person had received the highest dosage.

Carlton also began working with a major European cheese company to develop a strain of bacteriophage to kill listeria, a bacterium that can contaminate soft cheeses like brie and camembert. Says Carlton, "The cheese processors can't use antibiotics, because they would also kill off all the healthy bacteria used for ripening." Each year, a few dozen children in France and Germany die from listeria poisoning – a condition that results in severe diarrhea and can develop into meningitis. In the U.S. pregnant women are advised to avoid soft cheeses because of the risk of listeria poisoning, which can cross the placenta. While such contamination is not common, it's a perception problem that cheese companies are eager to eradicate.

Does Carlton believe that phage therapy will one day replace antibiotics? Not at all. "I see a scenario where doctors will use phage therapy and antibiotics," said Carlton over a sushi lunch not far from his office in Port Washington. "Because there's almost no chance that a bacterium can resist both phages *and* antibiotics. Say you have a patient with a Methicillin-susceptible strain of staph. If a doctor gives Methicillin, the patient may develop strains that are meth-resistant. How can you stop a strain of staph from becoming resistant? If you give the patient phages and methicillin, the bacteria are not likely to be able to become resistant to both, because the mechanisms of resistance are fundamentally different and thus resistance to one agent will not enable the bacterium to be resistant to the other. Phages will be used in combination with antibiotics. That's the way phage therapy should be practiced."

Chapter 8
The Startups

Superbugs in Baltimore

In the days when Merril and Carlton were reading about antibiotic resistance, a physician in Baltimore, Maryland was experiencing it as a daily reality. J. Glenn Morris, then head of infectious diseases at the University of Maryland's Baltimore Veterans Affairs Medical Center, was one of the first doctors in the country to recognize the emergence of VRE in the late 1980s. Morris, now director of the Emerging Pathogens Institute at the University of Florida, Gainesville, was then splitting his time between lecturing at the University of Maryland's medical school, where he was a professor of epidemiology, and making hospital rounds. He began noticing that a growing number of his most seriously ill patients were coming down with infections that didn't respond to medication. At first, Morris wasn't surprised. The resistant bug was infecting mostly elderly patients who had been on vancomycin for extended periods of time. But soon, the number of cases became alarming. Between 1989 and 1993 the bug's rate of resistance to drugs grew from 0.3% to 7.9% nationwide.[1] When Morris's hospital undertook a systematic study of patients, he was shocked by the results: as many as one-fifth were carrying VRE in their guts. "It was sort of like a time bomb waiting to go off," he says.

In 1995, Morris met a patient whom he'd remember for years to come. He was an accountant in his early 40s with leukemia. While undergoing chemotherapy, the accountant had developed an infection that his body, with its reduced number of white-blood cells, was having trouble fighting off.[2] "When someone gets cancer chemotherapy, they tend to become infected with whatever is present in the

---

[1] Nosocomial enterococci resistant to vancomycin – United States, 1989–1993. MMWR Morbi Mortal Wkly Rep. 1993; 42:597–9, cited in J. Glenn Morris et al, "Enterococci Resistant to Multiple Antimicrobial Agents, Including Vancomycin, Annals of Internal Medicine, 15 August 1995, volume 123 issue 4, pgs 250–259.

[2] Michael Shnayerson and Mark Plotkin, "The Killers Within," (New York: Little, Brown and Company, 2002), pg. 44.

A. Kuchment, *The Forgotten Cure: The Past and Future of Phage Therapy*,
DOI 10.1007/978-1-4614-0251-0_8, © Springer Science+Business Media, LLC 2012

intestinal tract," says Morris. Had the bacteria been ordinary *Enterococci faecium*, doctors would have simply prescribed antibiotics and, within hours, the patient's symptoms – a high fever and chills – would have been a distant memory. But in this case, the patient had picked up a dangerous VRE strain from the hospital environment.

After examining the patient, Morris explained that there was nothing that could be done. However, once the chemotherapy was stopped, the patient's own body would step in and destroy the superbugs in his system. And that's exactly what happened – at first. The patient was sent home but, a few months later, a relapse of cancer forced him to resume chemotherapy. Once his immune system grew depleted, the VRE multiplied once more, and when chemotherapy was halted, his white blood cells regained the upper hand. Possibly, if the leukemia had not returned for a third battle, the accountant would have survived. But the cancer did return, and, with it, the superbugs. Eventually, the patient succumbed not to the cancer but to the bacteria. "As a clinician," says Morris, "it's very frustrating to stand by a patient's bedside and know you have absolutely no therapy to help them and basically to watch your patient die from a treatable infection."[3]

Around this time, Morris vented some of his frustration to a group of associates. It was 1995, shortly after the young leukemia patient had died, and Morris was talking to some of the fellows at the medical center about how to deal with antibiotic-resistant organisms. "Why don't you use phage therapy?" one of them suggested. The room fell silent, and everyone turned to look at the round-faced microbiologist who had recently arrived from Tbilisi, Georgia. "I just gave him a blank stare," says Morris. "What was phage therapy?"

Alexander Sulakvelidze, then 31, was on a U.S. National Academy of Sciences grant, awarded to citizens of the former Soviet Union, to study how certain types of bacteria cause disease in humans. Over the course of his many months in Baltimore, he and Morris had grown to be close friends. When Sulakvelidze, known to friends as "Sandro," arrived from war-torn Georgia in late 1993, Morris and his family cared for him the way a host family might help an exchange student, patiently walking him through the unfamiliar transactions of American daily life. Sulakvelidze had never before held a bank account, had never written a check or used a credit card. "Getting money out of an ATM machine was a science for me," he says. "Glenn and his family did more for me and my family than probably anyone else in this world."

Meanwhile, Sulakvelidze was just as shocked as Morris after their conversation. The Georgian scientist had grown up drinking bacteriophages for stomach ailments, watching physicians apply them to cuts and burns to fend off infections and knowing that phages could sometimes work in cases where other treatments had failed. He was stunned to learn that it was not part of the American arsenal against infections. "What evolved," says Morris, "was the interesting process of going back in time and looking at phage therapy and trying to get a feel for what its

---

[3] Author interview with Morris in Baltimore, October, 2001.

potential utility is in an era where suddenly we are beginning to realize that antibiotics are not always the answer."

Sulakvelidze's dissertation advisor in Tbilisi had been Teimuraz Chanishvili, then the scientific director of the Eliava Institute. Sandro called him up to see who at the institute might be interested in applying for a collaborative grant to look into the possibility of using phage therapy in the West. Chanishvili put Sandro in touch with his niece, Nino Chanishvili, who headed a group of researchers at the institute. Together, they applied for a grant to the Civilian Research and Development Foundation, a nonprofit established by the National Science Foundation to support international collaboration among scientists.

The Millionaire and His Mission

One year later, in November 1996, Caisey Harlingten sat flipping through the latest issue of Discover magazine. Seated next to his girlfriend on a flight from Toronto to Seattle, he had saved one article for last. It was called "The Good Virus," and in it, author Peter Radetsky described the work of the Eliava Institute and of the American scientists who had recently grown interested in the field. "I suggest caution until we can see some data," Bruce Levin, a population biologist at Emory University, told Radetsky. "It is definitely time for outside scientists to go [to Tbilisi] and take a serious look at what they've been doing." Reading those lines, Harlingten knew he had found his next mission.

Harlingten, then 36, was a stock broker-turned venture capitalist. He had made millions of dollars funding start-up technology companies and taking them public. In 1992, he'd helped found Microvision Inc., which specializes in a futuristic sounding display technology that enables images to be projected directly onto the retina of a viewer's eye. In 1995, he'd helped found and raise capital for a nanotechnology company based in Delaware.

Once in Seattle, Harlingten phoned Elizabeth Kutter, a phage biologist from Washington's Evergreen University, who had also been quoted in Radetsky's article. Kutter, another key player in the West's rediscovery of phage therapy, is a phage biologist. She had stumbled upon the Eliava Institute in 1990, while on an exchange organized by the Soviet and American Academies of Science. Based in Puschina, the Soviet Union's capital of biological research near Moscow, Kutter traveled to Tbilisi because she'd heard it was beautiful – and also to learn more about an institute that an acquaintance had told her specialized in phages. She was surprised to learn that its scientists used phages to treat infections. "I was skeptical at first," she says, "but the more I talked with people from the Institute, the more seriously I began to take it."

Kutter formed a deep bond with the scientists at Eliava and has served as the Institute's main bridge to the West. Each year, Kutter invites one or two Eliava graduate students to Evergreen to study and secures grants to bring Eliava scientists to her annual phage meetings, which used to cover only biology but now cover therapy as well. Kutter uses every opportunity to raise awareness in the West about the Institute's economic plight and vast stock of knowledge. Through a non-profit foundation she helped start, PhageBiotics, Kutter's family has given close to $100,000 to the institute in the last decade. Kutter told Harlingten that she would be happy to accompany him to Tbilisi and to introduce him around, just as she had done with Radetsky.

They left almost right away, in November of 1996. In Georgia, Harlingten grew even more excited about his nascent project. He met Nino Chanishvili, the niece of Institute Director Teimuraz, who promptly suggested that Caisey partner with her, Morris and Sulakvelidze. To Harlingten, it seemed a perfect match: he wanted to maintain contacts with Tbilisi yet work through a reputable lab in the United States.

Later that winter, Harlingten flew to Baltimore to meet with Morris and Sulakvelidze at their University of Maryland offices. "I'm expecting a millionaire to come walking in," remembers Sulakvelidze, "and in comes a guy with no luggage wearing sneakers and jeans. He speaks impressively, but at the same time I wondered: is this for real?" After some initial sizing up, Sulakvelidze quickly grew to like Harlingten, and they agreed on a partnership. Harlingten promised to fund an initial round of collaborative research between Morris, Sulakvelidze and Nino Chanishvili at $150,000. He also gave Chanishvili about $10,000 to set up a small lab with two research assistants in an abandoned building on the grounds of the Eliava Institute. Sandro and Glenn gathered 500 samples of VRE from deceased hospital patients in Baltimore and mailed them to Tbilisi, where Chanishvili set about matching them up with phages from her stock. The goal: to find a bug in Tbilisi that could save lives in Baltimore.

The collaboration was a success. Nino found matching viruses, propagated them, and watched as they worked their magic in a test tube.[4]

Buoyed by these results, Harlingten formed a company called Georgia Research, Inc. based in Bothell, Washington and brought on board a business partner, Richard Honour, a microbiologist and biotech analyst with whom Harlingten had worked in the past. Then, having traveled to Poland and Canada to meet phage researchers there, Harlingten conceived of the idea of convening the first international congress in recent memory on phage therapy. In the spring of 1997, scientists from Germany, Poland, England and Canada, accompanied by a BBC camera crew, gathered at the former Communist Party headquarters in the mountains above Tbilisi to give research presentations and discuss the future of phage therapy. "I'm not sure where we are. We're in the wilderness," Harlingten excitedly told the BBC. "We feel like pioneers."[5] He had grown deeply fond of Georgia and of his new associates at the Institute. He could see that, for them, there was more at stake than a business venture: they were looking to Harlingten to rescue them from the brink of scientific extinction.

---

[4] From "The Virus That Cures," BBC Horizon series and author interviews with Alexander Sulakvelidze.

[5] "The Virus That Cures".

But Honour had other ideas. He took one look around the decrepit Eliava Institute labs and felt his friend was being naïve about the possibility of doing business there. "Richard Honour was apparently not very happy from the beginning, and he was not hiding it too well," remembers Sulakvelidze. "It was obvious that he didn't want to deal with Eliava and had his own agenda."

Still, some business negations did go on. The Tbilisians, led by Nino and her uncle Teimuraz, wanted Honour and Harlingten to invest $1 million to turn the Institute into their company's world manufacturing and distribution center. But Honour couldn't imagine the Food and Drug Administration approving any product that came out of there. He'd been particularly shaken by a large portrait of Lenin that still hung in the lobby back then: it hid damage from a mortar round and was itself pockmarked by bullet holes. "People there are nice," he said, "but it's a truly lawless society. I said, 'Caisey, this is very interesting but there is no way on earth that any of their very old phage materials could be brought forward.'" In the States, he figured, they'd be able to manufacture the same phages for very little money – and would have better control over quality. After days of talks, Honour and Harlingten announced they were heading home, but didn't definitively break off the deal – Honour was afraid of the consequences. "I told Caisey, 'We have to get out of here, we're gonna get killed. Which way's the airport?' And we boogied out of there for London."[6]

Back home, Harlingten and Honour wrote up a contract and faxed it to Nino and her uncle. It offered the institute $75,000 per year for 2 years, plus 2% of profits, in exchange for exclusive access to the Institute's phage therapy expertise and to any and all resulting products. The institute refused to sign. In the end, Honour and Harlingten's visit created a bitter rift among Eliava scientists. Many accused Nino of greedily controlling – and then bungling – the talks with Harlingten. Chanishvili, for her part, blamed the BBC. The program, she said, announced that talks had broken down before they really had.[7] In a New York Times article, she also accused Harlingten of swindling the Institute. "We gave the Americans access to all this background research, and they simply walked away with it," she said. "They told us we were stupid in business. Well, that at least was true."[8]

Talks between Sandro, Glenn, Caisey and Richard broke down soon afterward. "After the summer," says Sulakvelidze, "there were some strange emails from Richard and Caisey. Caisey clearly was torn. His e-mails were very confused. And it was obvious to me that he was sort of figuring out between what he was told by Richard and what he wanted to do and what he was hearing from Nino and us. So it was a hard time for him." Honour was insisting that any animal studies the company performed would have to be done in Bothell under the supervision of scientists more familiar than Sulakvelidze and Morris with the FDA regulatory process. Sulakvelidze

---

[6] Author interview with Honour, 07/05/02.

[7] Author interview with Chanishvili 11/07/02.

[8] Osborne, Lawrence, "A Stalinist Antibiotic Alternative," Feb. 6, 2000. The New York Times Sunday Magazine.

wanted to do at least the preliminary work in Baltimore. "I did not like Richard's heavy-handed approach," said Sulakvelidze. "I didn't like the fact that we had started out with Nino and Eliava and suddenly they got dumped. So pretty much we chatted, Glenn and I, and agreed we didn't want to do anything with that company." Instead, they began exploring the possibility of going into business on their own.

Meanwhile, Honour and Harlingten changed the name of their company to Phage Therapeutics and began examining the possibility of genetically engineering phages to attack a broader spectrum of bacteria than was possible in the wild.

Looking back on the incident years later, Harlingten says he should have known better than to invest any money in Tbilisi (in the end, he estimates, he sank about $150,000 into his collaboration with Eliava). Betty Kutter, he says, took him there so he could "get indoctrinated by the Georgians that they had something special." In fact, the place was nothing but "a memory, an echo from the distant past." Says Kutter, "I'm embarrassed that I brought him there."

A Midnight Epiphany

This mid-1990s burst of Western media interest would bring one more company into the race to reestablish phage therapy. Sitting in his newly built house in Bangalore, India, Janakiraman Ramachandran flipped on the TV and happened to catch a rerun of the BBC documentary filmed around Harlingten's summit: "The Virus That Cures." The program made such a deep impression on Ramachandran, who was then president of the Indian research and development unit of the Swedish pharmaceutical company AstraZeneca, that he still remembers the exact date when it aired: April 10, 2000. In just 2 months, Ramachandran would be retiring from AstraZeneca, having reached the maximum age of 65. He had devoted the past 10 years of his life to harnessing first-world technology in the fight against infectious disease in the developing world. In particular, he had spearheaded the company's ambitious efforts to develop a new drug for tuberculosis. Now, suddenly, his TV was telling him about a seemingly inexpensive treatment that sounded like a miracle cure for almost any bacterial infection. (Infectious diseases account for 90% of the developing world's health burden, and the vast majority of these infections are caused by bacteria.)

He phoned his friend and longtime colleague, Anand Kumar, then head of AstraZeneca's non-profit research foundation: "Are you watching the Horizon series?" Ramachandran asked.

"No, it's midnight."

He suggested Kumar catch the rerun at 2 am.[9]

---

[9] Author interview with Anand Kumar, 01/06/04.

Ramachandran was born in nearby Madras, now Chennai, and earned his Master's and Ph.D. in biochemistry at the University of California in Berkeley. After stints at Israel's Weizmann Institute of Science, near Tel Aviv, and Bangalore's Indian Institute of Science, he served on the faculty of the University of California, San Francisco. From there, Genentech, the San Francisco biotechnology company known for its portfolio of cancer drugs, hired him away in 1984. When Astra, before it merged with Zeneca, offered him the directorship of a new non-profit foundation that would develop drugs against third-world diseases like malaria and cholera, he jumped at the chance to help the country where he was born.

After watching "The Virus That Cures," Ram went on the Internet to find out more about phage therapy. He quickly found Betty Kutter's Web page at Evergreen University, where she had posted a lengthy article on the history of phage therapy. From that paper, Ramachandran learned for the first time about an even earlier phage discovery made in his own homeland. In 1896, M. E. Hankin, an English scientist working at the government lab in Agra, India, had set out to discover the source of the epidemics of cholera that periodically swept through central India, particularly during mass religious pilgrimages to the banks of the Ganges and Jamuna rivers. European authorities, having watched thousands of inhabitants washing themselves, their cattle and their clothes in the cloudy waters – and also dumping partially burnt corpses into them – were convinced that the Ganges and Jamuna were breeding grounds for the germs. Hankin set out to determine if this was so. Noting a presence of cholera microbes in the water surrounding bathers and spots where corpses of cholera victims had recently been deposited, Hankin wondered why cholera epidemics never flowed downstream. "The basic law concerning the progress of widespread epidemics in India is that, having originated in Bengal, they go upstream the river and its tributaries," he noted. The bacteria, he postulated, must be unable to survive in the water – perhaps because the Ganges and Jamuna lacked the nutritional matter to support the bacteria. Instead, his experiments led him to the startling conclusion that "these waters contained an antiseptic that had a powerful bactericidal action on the cholera germ." In other words, water from the Ganges river, when mixed with cholera germs in a test tube, killed the bacteria in less than 3 hours flat. Something in the water destroyed the germs before they could travel downstream and infect other communities. Conducting similar experiments with tap water and well water, he found that drinking water did not kill the germs and that well water, in fact, was "a good medium for this microbe." Based on this finding, he recommended that, "during Hindu pilgrimages to sacred places on the banks of the Ganga and the Jamuna, it would be better to advise against the use of well water and encourage the use of river water."

Today, Hankin's paper on the subject is cited as the first observation of phage action – although Hankin himself never discovered what it was that accounted for the water's bactericidal properties. "Although the scientific interest of the preceding results may be limited by the fact that I have not yet discovered the nature and origin of the antiseptic substance present in the waters of the Ganga and Jamuna rivers, what appears interesting is that they explain why cholera does not travel downstream rivers in India," he concluded.[10]

To Ramachandran, the experiment tied phages in with a long held belief in India that the Ganges river has sacred, curative powers. Households across the country keep a small copper vessel filled with water from the Ganges to preserve their luck and well-being. Around the same time, William Summers's biography of Felix d'Herelle was published and in the pages of that book Ramachandran learned about another Indian connection: that d'Herelle had traveled to India himself, and that the British government, lead by the Haffkine Institute's J. Morison, had carried out convincing studies that showed its effectiveness. "That was the final attraction," he said. "Unlike other biotech approaches, where in many cases you spend 2 years figuring out if your concept will work and if it's a good idea, here we didn't have any concept to prove: we knew that if you get the right phage to the pathogen, it will work." The Morison studies especially impressed Ramachandran, who points out that it was basically a controlled study: in the villages that accepted phage, 10 people died of cholera per year, while the districts that had refused the prophylactic treatment lost more than 1,500 inhabitants to raging epidemics. "Nobody mentions this," he says. "Nobody looks at this data, and it's very solid data."

In August, after his retirement, Ramachandran flew to Washington to visit Kutter. Evergreen University was not a long haul for Ramachandran, who has a house in Palo Alto where he spends about half his time. A few months later, Ramachandran's colleague, Dr. Kumar, the recipient of his midnight phone call, invited Kutter to Bangalore where she gave a lengthy presentation on phage therapy in honor of Ramachandran's retirement. Kumar had the speech printed and bound, and copies of it are still kept in the AstraZeneca research library.

Upon returning to Bangalore, Ramachandran invested $100,000 over a period of several months to launch his own phage therapy company. He called it GangaGen, for "Ganga Genesis," or originating from the Ganges, and incorporated it in Bangalore in September 2000. "I decided to set it up in India because of the advantages of cost, and because there were plenty of people available to be recruited." Bangalore is India's high-tech capital, having first attracted attention when the American computer company Texas Instruments relocated its R&D center here in the mid 1980s. Since then, a host of other high-tech companies, including IBM have followed suit. All take advantage of India's elite technical and science universities, like the Indian Institute of Technology and the Indian Institute of Science, with campuses in major cities across the country.

Ramachandran rented 800 square feet of office space not far from downtown Bangalore. At that point, he felt, it was too early to seek investors. "There was no intellectual property," he said. "Without that you can't do anything." So, his first step was to begin collecting a library of phages with which to work. He hired two

[10] Hankin, M.E. "The bactericidal action of the waters of the Jamuna and Ganga rivers on cholera microbes." Translated from the original articles published in French, Ann. De l'Inst. Pasteur. 10.511 (1896).

staff scientists and about five or six research assistants, whose job was to contact local hospitals and collect sewage samples from them. "We were following d'Herelle's principal that in a hospital there is always somebody naturally recovering and therefore there is a chance they've been exposed to the phage in the environment. In fact, that proved to be true," says Ramachandran.

Soon, like Harlingten, Ramachandran made up his mind to travel to Tbilisi and hired two scientists to come to Bangalore for 2 weeks to help isolate phages. Though both sides hoped the collaboration might lead to something more long-term, it wasn't to be. "My main interest was to see if we could get a head start by working with them," explains Ramachandran. "Unfortunately, at that time, [the Eliava Institute] had really deteriorated and half the place didn't have electricity and light. And I didn't have confidence that there was any leadership there that could guarantee that any cooperation would work." The break with Ramachandran was another disappointment for the Tbilisi scientists, who were still hoping to find a lifeline in the West.

Meanwhile, even though Ramachandran wasn't seeking investors, they managed to find him. One day, a colleague of Ramachandran's asked him to help two venture capitalists perform due diligence on a company in which they were considering investing. When he met the VCs, Norman Prouty and Vijay Angadi of Bangalore-based ICF Ventures, they asked what Ramachandran was working on. Soon, they were so intrigued by phages that they began reconsidering their original plans. "They came back the next day and spent another 2 hours talking about phages," recalls Ramachandran.

A few days later, they told Ram they were ready to invest $2 million, as long as he agreed to also incorporate in the United States, to make it easier for their American investors to receive their returns. Ramachandran incorporated in Delaware, and his company closed its first round of financing Oct. 31, 2001, although it still had no intellectual property to its name. "Norman told me he invests in people, not corporations," says Ramachandran.

Chapter 9
Four Companies, Four Strategies

"I'm here to talk about one of the biggest surprises of my life."

Alexander Sulakvelidze was standing behind a podium in the lecture hall of a Bethesda, Maryland hotel in July 2003. "Don't worry," he added, with a mild Georgian accent, "it's nothing too personal."

Sulakvelidze, whose round frame, broad face and black beard streaked with white give him a distinct panda-like appearance, prompted the technician to turn down the lights and launch the slide show. "Growing up in Georgia," explained the microbiologist, "I had always looked at phage therapy as a given. It was always right there, as an alternative to antibiotics. Only after coming to the United States did I realize it wasn't available here." The screen flickered to life with a grainy video of Stalin presiding over a Kremlin military parade and then of wounded Red Army soldiers in World War II who had packed vials of phage in their First Aid kits. Sulakvelidze spoke about the history of phage therapy in the Soviet Union and about Intralytix, the name he and Morris eventually gave to the company they founded in 1998. "We live in a sea of bacteriophage," he concluded. "The key is to find the right ones and use them in the most effective way."

Sulakvelidze was speaking at an annual conference on drug resistance, sponsored by the National Foundation for Infectious Diseases. He had been giving this same talk for the past 6 years, using it to enlighten potential investors, regulators, funding agencies, pharmaceutical companies, collaborators and journalists. Standing before a room full of people, speaking a foreign language, Sulakvelidze looked completely at ease. And yet, just 10 years earlier, he would never have thought he'd leave the country where he was born.

In 1992, Sulakvelidze was a newly minted Ph.D. working in Tbilisi. At 27, he had become the founding director of the first division of molecular biology at Georgia's National Center for Disease Control and was well on his way to a distinguished scientific career. His mission was to establish molecular biology, the study of the tiny components that make up living cells, as a mainstay of Georgian science. (Under the Soviet Union, molecular biological research was carried out in Moscow and in the Russian city of Puschina; after Georgia declared independence from the USSR in 1990, the new country found itself bereft of scientific institutions.)

A. Kuchment, *The Forgotten Cure: The Past and Future of Phage Therapy*, DOI 10.1007/978-1-4614-0251-0_9, © Springer Science+Business Media, LLC 2012

From his office, decorated with a fancy Turkish rug and equipped with three telephones – a sign of status – but no computer, Sulakvelidze thought of all the projects he would like to undertake in his spacious new lab, which occupied an entire floor of a brand new six-story building. But, because Georgia lacked the most basic equipment for molecular biology, each experiment required elaborate arrangements, including hunting for enzymes at other labs and traveling to Moscow. "I got increasingly frustrated," he said. "Once you have an idea for an experiment, an idea to test, it becomes almost unbearable to have to wait a month and then to go to Moscow to be able to test it. I just couldn't do anything where I was."

The scientific arrangements were complicated by the chaotic state of Georgia's politics. In 1991, just one year after the country declared independence, a bloody civil war engulfed the capital. A large segment of the population had grown dissatisfied with the country's first elected leader, Zviad Gamsakhurdia, and, together with members of Georgia's armed forces, launched a violent coup. Fighting broke out in the streets, as rebels and pro-government forces clashed outside the House of Parliament on Georgia's Rustaveli boulevard, eventually leading to Gamsakhurdia's flight and replacement by former Soviet foreign minister Eduard Shevardnadze. All around Sulakvelidze, in the center of the elegant city where he had grown up, stray bullets crashed through apartment windows, buildings burned, and loud explosions went off throughout the night. There was no public transportation, electricity, hot water or heat. "Who thinks, at those times, about science?" says Sulakvelidze.

Sulakvelidze had never before thought about leaving his country, especially since becoming engaged to a young journalist he'd met in college. "Immigrating was never an option," he says. "I could never in my life imagine that I could live outside of Georgia. The psychology [there] is a little different. In most cases in those years, you were born in Tbilisi, you would grow up in Tbilisi, go to school, go to college, go to work, and you would actually die in Tbilisi. So, all your friends would be there, all your relatives. It's a different society, and, in my mind, I was very much sitting in that system." Yet, this time, he realized he would have to make a change, to do something "out of the box."

That was when he applied for, and won, the National Research Council Fellowship that brought him to Baltimore. Sulakvelidze had sent out about a dozen applications – all typed on his manual typewriter. Morris had been the most receptive toward hosting him because he had been studying the same bacteria in which Sulakvelidze was then interested: *Yersinia enterocolitica*, which causes food poisoning and is a less lethal cousin of the Bubonic Plague. Realizing that the deadline for the fellowship was fast approaching (in those days mail took 45 days to travel from the United States to Georgia), Morris took the liberty of retooling Sulakvelidze's project outline, added his own supporting material and mailed it back to Sulakvelidze for him to sign and submit. "He was a very bright candidate with a strong background – and also somewhat persistent," says Morris. "Bringing scientists over is always a gamble. But he turned out to be a brilliant scientist."

Sulakvelidze initially came over for 9 months. But, when his research began to yield interesting results – he was comparing Georgian and American strains of Yersinia in an effort to determine how they cause disease – Morris encouraged him to extend the terms of his stay. Sulakvelidze returned to Georgia, married, and brought his new wife back with him to the United States.

***

The first time I met Sulakvelidze and Morris, in the fall of 2001, they were running Intralytix out of their offices at the University of Maryland. Morris, who is stocky, with a salt and pepper beard and mustache, has an abrupt, sober manner that contrasts sharply with his business partner's. Sulakvelidze is easygoing, warm and almost always smiling, and it's occurred to me more than once how easily Sulakvelidze, who grew up under communism, has made the switch to running a biotech startup in the United States.

Sulakvelidze showed me into his office, in a drab highrise overlooking the University of Maryland campus, and launched the same slide presentation I'd see 2 years later at the NFID conference. Sandro explained that Intralytix was developing phages for use against human disease as well as against bacteria that contaminate fruits, vegetables, eggs and meat. "Today there is no consensus in this country that phage therapy is effective," he said. "You can find a sharp division in the scientific community between believers and skeptics, so I think what needs to be done is rigorous pre-clinical and clinical trials of phages." He and Morris were working on developing phage cocktails – mixtures of five or six different types of phages that could be used against a variety of bacteria.

VRE, they explained, was their main target when it came to human diseases, and they were preparing to petition to the Food and Drug and Administration for permission to launch preclinical trials. "VRE is a real bear," Morris chimed in. He had arrived after Sandro started the slide show. "We basically had 50 years during which time we really could kill just about any bacteria we came across. This was the antibiotic era, it was the golden age of medicine. And the door is at least beginning to close on that era." Today, there are two new antibiotics used to treat VRE, but they have serious side-effects and the bacteria are already growing resistant to them.

The other pathogen that Morris and Sulakvelidze were interested in was *Staphylococcus aureus*, the germ that nearly cost former talk-show-host Rosie O'Donnell her hand earlier that same year and had also plagued Fred Bledsoe. O'Donnell was so traumatized by her infection that she wrote about the ordeal in her now defunct magazine. The cover image was of O'Donnell brandishing her bandaged left hand – wrapped so many times that it resembled a white boxing glove. In contrast to enterrococci, staph aureus is highly virulent, and can infect patients who are otherwise completely healthy. "VRE mostly effects the immuno-suppressed, so the damage of having this superbug was limited," said Morris. "In contrast, if we indeed develop a super staph aureus, then we are going to have very major problems. A lot of our ability to do work in medicine today is dependent on our ability to kill it." Found on the skin, staph causes infections when it penetrates wounds and enters the bloodstream. On the simplest level, it causes boils, but when it enters the blood it can lead to bacteremia and heart-valve infections. Today, isolated cases of Vancomycin-resistant staph aureus have already appeared and, says, Morris, "It's only a matter of time" before there are more.

After Sulakvelidze and Morris's break with Harlingten and Honour, they continued working with the Eliava Institute, but that collaboration too eventually broke down. "I think there are good people at the institute whom I wanted to continue funding, but then the financials kicked in, so we just couldn't afford to do it," he says. "But the truth is that no other company has funded them either, and that is because it just would take more money and more efforts than doing it here. And, plus, there will always be an element of the uncontrollable: the Georgian legislature, the infrastructure, who belongs to whom. There are a lot of unknowns, and I don't think companies will [risk] that."

***

A few months after I met with Sandro and Glenn for the first time, they had raised enough money to move into a spacious, high-tech lab on the Baltimore waterfront. Its location, on a pier next to an old Coast Guard ship from World War II is prime, because Intralytix gets most of its phages from the harbor that surrounds it.

One day, Roy Voelker, then Intralyx's director of food safety, showed me how he fishes for phages in the Chesapeake Bay. Our mission was to find a phage for *Clostridium perfringens*, a bacteria that kills chickens and is a major cause of food poisoning from undercooked poultry. Inside the lab, Voelker picked up a long orange rod to which he'd jerry-rigged a plastic container, and we headed down to a brick, tree-lined walkway that runs along the water. He knelt down on the edge and dipped the jug into the harbor. From experience, he knew the location of three pipes that spouted sewage and industrial waste, and that's where the phages would be. Once the jug was filled with brownish water, we headed back upstairs. "We used to look for phages in sewers on the street," he said, by removing manhole lids or drain covers. But passersby would stop to ask what they were looking for, and when Roy replied "viruses," they looked alarmed.

Back in the lab, Voelker emptied the contents of the jug, through a filter, into a sterile, glass flask. The filter would keep out bacteria but let through any phages that might be present. Next, in a separate container, he mixed powder with water to create a meat broth for growing bacteria. From a nearby refrigerator, he took out a small glass vial marked CPK7, for *Clostridium perfringens*, scooped out some clotted bacteria using a wand with a loop on one end, and added it to the broth. In the final step, he took a plastic dropper, filled it with 5 mL of the bacteria mixture and dropped it into the flask containing the harbor water, then moved the entire mixture into a heated compartment where it would incubate overnight while the phage and bacteria multiplied. The next day, the lab would separate the phages from the bacteria by spinning the mixture at 8,000 times the force of gravity in a centrifuge, running it through a filter once more, then adding the filtrate to a lawn of clostridium bacteria grown on a Petri dish to see if any of d'Herelle's *taches vierges* formed. If they did, the phage would be characterized – i.e., its DNA would be sequenced – and it would be added to Intralytix's library of some 600 different bacterial viruses.

***

It was in the spring of 2002 that Intralytix first began petitioning the Food and Drug Administration for permission to launch preclinical trials of phage therapy for use against Vancomycin Resistant Enterococci. Using phages first isolated in Tbilisi, Intralyx's lab workers succeeded in rescuing mice infected with VRE. Sandro used these results in his lengthy Investigational New Drug permit application, or IND.

By this time, Intralytix had assembled what it felt was an impressive team of experts to help it through the daunting process of regulatory approval. Its chairman of the board a tall, ruddy-faced M.D. named Torrey Brown, once served as Maryland's State Secretary for Natural Resources, was assistant dean at the Johns Hopkins School of Medicine, and had also headed a non-profit agency that specialized in helping small firms conduct clinical trials. Intralytix also retained an outside consulting company that specialized in steering biotech startups through the FDA's regulatory maze.

But perhaps Intralytix's biggest ally was Marissa Miller, an antimicrobial resistance program officer with the NIH's National Institute for Allergies and Infectious Diseases. Miller has helped Intralytix apply for government grants and has coached them through their application to the FDA. "We're hoping that phage will provide a new avenue of therapy in these very difficult multi-drug resistant bacterial infections," she said. In 1999, Congress had given the NIH an additional $20 million to partner with companies in an effort to stimulate the development of new drugs to battle antibiotic-resistant infections. "We're looking at emerging infectious diseases where there's a high public health need but low industry interest," she explained. "We're trying to stimulate companies to come into areas where there may not be a really clear profit motive."

Over the last 20 years, major pharmaceutical companies have all but walked away from infectious diseases. Even as the number if antibiotic resistant infections in the United States continues to climb, the number of new antimicrobial agents approved for sale has fallen every year from 1983 through 2007. As the U.S. population ages, the medicines that have come to promise the largest financial returns are those that target chronic conditions like high cholesterol, hypertension and heart disease. "Antibiotics are given for a short period of time: 4, 7 and 14 days," says Miller. "So those are not as profitable as products you give over a lifetime." Voelker, in fact, is a walking example of this. He worked for Smith, Kline, Beecham, helping develop new antibiotics, until they shut down his department in 2001. When that happened, he decided to return to an old interest he'd developed in graduate school, bacteriophages, and found Intralytix through a Google search.

As pharmaceutical companies have grown, their willingness to take risks has waned, and now many of the newest drugs come from biotech startups. For any new drug, the hurdles are overwhelming: by some estimates, as few as 18% of medicines that make it to Phase I clinical trials reach the market. And the process of drug development and approval costs an average of $800 million and takes 10 years. So, instead of developing drugs themselves, large pharmaceutical companies have preferred to wait out the risk and purchase the startups with the most promising products after they've advanced farther along the road to regulatory approval.

With phage therapy, says Miller, there are still plenty of hurdles to overcome, including the widespread skepticism on the part of regulators. "It's not met with a lot of support in this country," she says. "I'm not sure exactly why, except this country has been so drug oriented. I think that's why it's a niche for smaller companies." She added, "Every novel approach we can come up with, we need to pursue."

The NIAID told Sulakvelidze that they may be able to help him with a $200,000 government grant that would fund the initial stages of his pre-clinical work. Armed with the promise of financial support and numerous consultants, Sulakvelidze spent 1 year putting together an application and mailed it off. Several months later, the FDA called to schedule a conference call in which they would report their decision.

On June 10, 2001, Sulakvelidze climbed into his dark purple Camaro and drove the 40 miles to the NIH campus in Bethesda. There, he met Marissa Miller and Brown in a conference room, and they were patched through to Washington. Because bacteriophage therapy was new to the FDA, the government had convened regulators and experts from many different divisions. There were experts from the FDA's Center for Biologics Review, toxicology experts, microbiologists, and regulatory affairs specialists from the National Institutes of Health. From Intralytix, there was Sandro; Torrey Brown; Glenn Morris, who was on the phone from Baltimore; plus two University of Maryland physicians called in as ad hoc consultants.

The FDA's initial response to the IND was not what Sulakvelidze had hoped for. They peppered him and his staff with numerous questions, some of which he felt were reasonable and others not. One of their major concerns was that only lytic phages – the kind that destroy bacterial cells quickly and are less likely to exchange genes with other organisms – be used in Intralytix's final product. They felt Intralytix had not demonstrated conclusively that it had found a fool-proof way of distinguishing one type of phage from another. The government also suggested that Intralytix move to a different animal model: in the one they used, the mice did not die of VRE, nor were they immunocompromised. A different model might more closely resemble the types of patients who would be the best candidates for phage therapy: the sickest of the sick. What Sulakvelidze felt was a less reasonable request was that the FDA also asked him to determine the rate at which each of the six phages in Intralytix's proposed VRE cocktail would mutate inside an experimental animal – and how these mutations would effect the composition of that animal's gut flora. "That would be very difficult to achieve in any reasonable amount of time or with any reasonable resources," he said. "And, quite frankly, I wasn't sure that those things were really needed. My rationale was: look, we have these same phages on our skin and in our guts. Who is looking at their mutation rate? If they persist in this, it will be very difficult for us to carry on." The call ended with

Sulakvelidze saying they would go over all the points and get back to the FDA as soon as they could.

Walking out of the building, Brown told Sulakvelidze he felt that the meeting had been a big success. Sulakvelidze wished he could have shared his optimism. "What is good and what is bad?" Sulakvelidze told me a few weeks later. "To my mind, good is that it's beautiful, and everything behind that is sort of bad. I mean, the worst is probably that they would have said, 'Don't even consider this.' But short of that, they said almost everything else." Later in the conversation, he added, "I think the biggest problem was perhaps the lack of understanding that this technology will be very different from antibiotics. I can understand if there was a chemical formula, that you can spend 5 or 10 years analyzing it, and then you could use it for 20 years without changing its composition." Phage cocktails, on the other hand, would have to be altered and remixed after a year or two, like the flu vaccine, to account for bacterial evolution. If each company had to re-analyze the mutation rates of every new phage, the length of time involved and the cost would grow prohibitive.

Despite Sulakvelidze's initial disappointment, he convinced himself that he'd probably be able to find a middle ground with the FDA. Perhaps he would have to submit a thorough analysis of only the first phage cocktail to go on the market and then would be able to have greater flexibility with the subsequent ones. So, Sulakvelidze went over each of the FDA's points, which the agency mailed to him following the teleconference, and tried to determine how much money and time it would take for Intralytix to generate the necessary data to resubmit the application. He decided it would require about 2 years and close to $1 million, money the company did not have. But he hoped that the NIAID's promise of $200,000 would help them generate enough solid data that he'd be able to attract new investors with it. "It would have been a big step forward," Sulakvelidze told me, sitting behind his desk, a photo of his 4-year-old son Levan smiling out from the screen of his laptop. "But then congress cut or reallocated the budget of NIAID. So, the bottom line was that they did not have the funds to do it. We shelved the entire VRE project. It's frozen. We have a couple of phages in the refrigerator, we have the project outlines and study designs, but that's it. We do not work on it."

From this point on, Intralytix decided to focus on the use of phages in meat, poultry and agriculture, and to resume its human work only after the company began earning revenue.

***

Meanwhile, Ramachandran was still making good progress with his company in Bangalore. After ICF Ventures promised Ramachandran that it would invest money in his company, he turned to the challenge of developing patents. No longer was his own fortune the only one at stake – now the money of about a dozen people he'd never met was riding on his success. He flew to Palo Alto, immersed himself in the stacks of the Stanford University library and spent weeks poring over every paper and book he could find on phage therapy. "I took the approach not of a phage biologist, because I'm not one, but of a biochemist developing a product," he said.

He made a list of all the perceived limitations of phage therapy. One was the concern that phages, upon breaking apart a bacterial cell, would release harmful toxins into the patient's body. That could lead to septic shock. The other was the possibility of phages swapping genes with a bacterium and, in the process, changing a harmless bacterial cell into a pathogenic one. While most bacteriophages can be considered friendly, allied with humans in a common battle again bacteria, there are other phages that function more like double agents, turning against humans and partnering with otherwise benign bacteria to make them into killers. The good phages are called "virulent" or "lytic." They eviscerate bacteria quickly and efficiently, usually in the space of 20–25 minutes. The quisling phages, called "temperate" or "lysogenic," invade a bacterium but, instead of killing it right away, lie dormant within it and integrate their DNA with that of the host cell. That process, called transduction, can have devastating consequences for human health. Lysogenic phages have created some of our most deadly strains of bacteria, including *E. coli* 0,157 and those that cause cholera and diphtheria. "Phages probably killed one third of Europeans who came to North American in the seventeenth century," says Carl Merril referring to diphtheria, which results when the harmless corynebacteria that line our noses and throats are colonized by Beta phages. "Many of the bacteria we'll have to kill were made pathogenic by bacteriophages," says Ry Young, the phage biologist from Texas A&M University. "That's the irony of it." One of the challenges of phage therapy lies in making sure that you are using lytic, not lysogenic phages and avoiding ones that might carry toxin genes.

Then, there were the commercial drawbacks. Because the idea of using phages to cure infectious disease is nearly 100 years old, the concept is unpatentable. Even if GangaGen were to isolate and characterize its own unique set of phages, the population of bacterial viruses is so enormous and diverse that another company could easily come along, find a slightly different virus that targets the same bacterial strain and legally be allowed to put it on the market as their own. And, finally, there was the problem of piracy. Says Ram, "We could put a product on the market, somebody could buy this product, open a garage fermenter, produce more of it and sell it. There is not much we can do, especially in countries where there is not a strict enforcement of some of these things."

Having finished his exhaustive survey of phage literature, he sat back and reviewed all the obstacles. "I said, 'Okay, how can we deal with these things?' And that's when I started looking at the way phages kill bacteria." Ramachandran had remembered something that most scientists overlook. While it takes a full 25 minutes for a phage to attack a bacterial cell, multiply inside it and burst out, the phage kills the cell during the first 5 min. If Ramachandran could figure out a way of preventing the phage from exiting the bacterium, he might overstep a number of hurdles. Without a "burst," no toxins or phages would be released into the patient's body. The fewer the number of phages, the lower the chance that a virus would swap genes with a bacterial cell. And, finally, if he could come up with a way of making all this happen, he would have solid intellectual property.

Ramachandran thought he knew just how to pull this off. Phages, as they replicate within a bacterial cell, release a protein called endolysin that eats away the inside of the bacterial cell wall, weakening it to the point where the phages can break out. If Ramachandran could block the gene that codes for that protein, he would have exactly what he wanted: a phage that kills its prey but never escapes.

Ramachandran filed a provisional patent application in September 2001, which gave him 1 year to develop his idea. He then called the husband and wife team of Sriram and Bharati Padmanabhan, two young molecular biologists who had worked for him at AstraZeneca and asked them to join GangaGen. They agreed and soon began putting Ram's idea to the test. Sriram took a T4 phage, the one with the most studied and best understood genetic structure, identified its gene for endolysin and plotted how to block it. He developed a plasmid, an artificial cell designed to swap genetic material with the bacteriophage. It would trick the T4 phage into giving up some of its genetic material and replacing it with a piece of DNA from the plasmid that would inactivate the endolysin. After Sriram created several of these mutant bacteriophages, his wife, Bharathi, tested them on a lawn of *E. coli* that she'd grown in a Petri dish. As the scientists expected, the phages didn't form any plaques, but when Bharathi tested the *E. coli* bacteria, she found that they were inactive: the phages had done their job beautifully but invisibly.

This approach would also help Ramachandran guard against piracy. In the first place, his phages would now be very difficult to reproduce – you would need more than just a garage fermenter to grow phages that don't burst out of their host. And, in the second place, he now had a way of marking his product. In place of the lysin gene, Ramachandran planned to insert a "signature GangaGen gene." Much the way DeBeers diamonds carry their company name in microscopic lettering, so too would GangaGen phages carry a tiny gene manufactured in their Bangalore lab.

The exciting technology had yet another practical application. Ram soon realized that scientists could use this method to develop vaccines. Today, vaccines are made with bacteria that have been killed with a formaldehyde solution or intense heat. "It's very harsh," says Ramachandran. "The heat inactivation destroys the determinants that produce the immune response." Vaccines work because bacteria, dead or alive, contain antigens – proteins on their surface that elicit an immune response. Patients form antibodies to these proteins that later protect the body from infection by these same agents. Ramachandran theorized that his technology would produce an organic vaccine that would leave more antigens unharmed on the bacterial surface. "It's a novel way of making a vaccine by killing from the inside," he says. "We are producing a dead bacterium that looks very much like a live bacterium." Bharathi tested the vaccine by administering it to mice and then injecting them with *E. coli* germs. Of the mice that were not vaccinated, 80% perished. But of the vaccinated group, all survived.

Because GangaGen was hoping to focus solely on phage therapy, it was looking to license this technology to a vaccine company that would be able to take it the rest of the way toward regulatory approval. Excited by these results, Ramachandran filed full patent applications for both products with the U.S. Patent Office on Sept.

27, 2002 – one year to the day after he'd filed the provisional patent – and GangaGen appeared to have left the harbor with the wind in its sails.

***

In early 2002, Ramachandran had a chance run-in with James Watson, the biologist and Phage Group graduate who, with Francis Crick and Maurice Wilkins, won the 1962 Nobel Prize for unraveling the structure of DNA. Watson was so intrigued by the idea of phage therapy that he offered up the Cold Spring Harbor Lab, the original home of the Phage Group, for a conference on the subject. Ramachandran, backed by ICF Ventures, invited some 30 scientists to the Banbury Center, Cold Spring Harbor Lab's private conference facility, to discuss prospects for the treatment.

On Nov. 14, the scientists gathered in a squat, modern cottage on the bucolic, 50-acre estate that makes up the Banbury Center. It was a sunny, crisp fall day, and a blanket of bright yellow leaves covered the grounds' luxurious lawns and still verdant shrubbery. When I walked in on that Thursday morning, delayed by my off-peak train from Manhattan, the conference room had already been darkened for the overhead projector, and the only figure I could make out was James Watson's. Sitting in the back of the room by the door, he wore khaki pants, loafers and socks that were slouched down around his ankles – and a trademark grin.

Ramachandran gave the opening speech: "It is highly fitting that Cold Spring Harbor Laboratory is the venue for the first scientific conference on Phage Therapy, since it was here that phage science originated six decades ago," he said. He went on to describe the global burden of infectious disease and the rapid emergence of drug resistant bacteria, then outlined the perceived barriers to the success of phage therapy: the bungled efforts of the 1920s and 1930s, the limited knowledge we have of any phages outside the seven studied by Delbruck's Phage Group. "Understanding and appreciating the enormous diversity of phage is essential to get us out of the mental straightjacket and one-glove-fits-all attitude that hampers progress in clinically relevant phage science," he said. "... We hope that this conference will launch a renaissance in phage science that will ultimately result in the approval and use of phages wherever and whenever appropriate."

With those words, the scientists launched 2 days of presentations and heated discussions about the future of this nascent field. Betty Kutter was there, as were Carl Merril; Sankar Adhya; William Summers, the Yale professor and author of Felix d'Herelle's biography; Ry Young; Marissa Miller from the NIH, plus specialists from England, France and Germany. Representing the Eliava Institute was Revaz Adamia, the head of a lab specializing in phage genetics who had been appointed Georgia's ambassador to the United Nations. He made the trip each day from Manhattan in a black Lincoln Town Car, turning the heads of all the other scientists. Adamia, a lively, affable figure who looks like a stockier version of Burt Reynolds from the 1970s, gave a talk on the last day about the history of his institute. Neither Caisey Harlingten, nor Sandro Sulakvelidze, nor anyone from any competing phage company was invited, and in the end, Ramachandran wound up hiring many of the attendees as staff or consultants for GangaGen.

Though the presentations were interesting, the most gripping moments of the conference came during the discussions that ended each session and during the lunch and dinner breaks. Ry Young, whom I met at the conference for the first time, stood out immediately. Wearing a navy blue baseball hat, a heavy brass belt buckle and running shoes, he had a loud southern drawl that echoed across the room.

After the first presentation, he called for a general renaissance in phage research. "The infrastructure needs to be built up," he said. "There has to be support for phage biology at the national funding agencies. There are lots of societal reasons for it, but phage biologists have really disappeared. The NIH and NSF [National Science Foundation] can fix this. It's so doable, it's so cheap." Miller, to whom Young's supplication was targeted, responded that the phage grant proposals she's received have not been of high enough quality to get funded. William Summers pointed out that there's been a general shift in science from an approach he called "How does the world work?" to one of, "What can I do with it?" In other words, there had been a shift from basic research to practical research that has direct applications in medicine. Another scientist, Vince Fischetti of Rockefeller University, suggested that phage therapists start by targeting the major pathogens. "That would really jump-start the field," he said. With that, the group broke for lunch.

We walked along a winding driveway, flanked by yellow and orange trees, toward Robertson House, the red brick colonial mansion where the scientists were staying. In the dining room, I sat with Betty Kutter and Gary Schoolnik, a professor of infectious diseases at the Stanford University Medical Center. It turned out that bacteriophage therapy had saved Schoolnik's mother from typhoid fever in 1947. "She'd been in the hospital for a month," said Schoolnik, who has silver hair, a quiet voice and a genial manner. Antibiotics for typhoid had not yet been developed. His father, a microbiologist-turned-surgeon, had read about phages that were effective against typhoid in a bacteriological journal. He contacted the authors of the study, a group of Los Angeles-based physicians, and had them ship their phages to Seattle, where the Schoolniks then lived. "My father injected the phage into her, and basically she responded immediately," said Schoolnik. Her fever dropped from 104 or 105 to normal within 48 h. "That was a very bold thing for my father to do," he said. "Today it would be illegal." After Gary met Ram, Schoolnik looked up the original phage therapy article that his father had read and mailed it to Ramachandran for his birthday.

After the next session, a major debate broke out about how to gain wider recognition for phage therapy. Mathew Waldor, a cholera expert from Boston's New England Medical Center said he thought everything about phage therapy reeked of nostalgia. "We have to move into the twenty-first century," he said. "The FDA wants exquisitely defined products." He alluded to an approach that Vince Fischetti had described at the conference, and around which Fischetti has since set up a private company. Fischetti is using phage enzymes, the chemicals that phages produce in order to enter and destroy bacteria, as antibiotics. To Waldor, it seemed that Fischetti's product was simpler and more predictable than an organism with its own DNA. To this, Kutter interjected that phages are useful precisely because they have their own DNA and can therefore make more of themselves as they work.

Suddenly, from the back of the room came a voice that no one had yet heard. It came from a diminutive Indian man who had dozed through many of the day's speeches. It was the founder of a biotech company that had developed and introduced a leading cancer drug. He said that the biggest challenge with phage therapy was going to be drumming up financial support. "You've got to sell this product," he said. "People, this stuff isn't sexy." He suggested the companies look into targeting some of the bacteria that can be used to make biological weapons, like anthrax and plague. Here, Ramachandran stepped in and urged people to not get ahead of themselves, thinking about regulatory issues and the future of phage therapy in the United States. "There is no clinical science on phages," he said. "That's what we should worry about." Miller added that she thought the companies should start by looking into environmental and agricultural uses of phage therapy – such as using them to kill foodborne pathogens – then gradually move into clinical applications from there. Within months, at least three of the companies would do just that.

***

By the fall of 2001, Caisey Harlingten had had a falling out with Richard Honour and his investors and abandoned the company to sail around the world on a yacht with his daughter. But the Bothell, Washington-based company, Phage Therapeutics, seemed to be going strong. In a phone interview Honour, then the president and CEO, described their ambitious plans to bioengineer superphages, each of which would attack a broad spectrum of bacteria. The phage he'd helped create, he said, had DNA that was three-times the size of an ordinary phage's genetic material. The added genes, he said, helped it beat back bacterial resistance, expanded its host range, gave it more stability and upgraded its shelf life. "It's a very unusual phage," he said. "We don't agree with what others do, which is isolating phages from sewers and wounds and rivers and use them until the host develops resistance. That's not the way real pharmaceuticals work." He also disagreed with Intralytix's approach of using phage cocktails, saying it would be too complicated to approach the FDA with a product containing multiple ingredients. "You have to concentrate on a single biological entity," he said. "Otherwise, you're just not going to get there."

Their lead product was an eye drop for staph infections of the cornea, a very small market but one in which he felt it would be easier to win FDA approval. "The FDA has asked us to stage the way we approach staph infections," he said. "We'll approach topical applications first, beginning with infections of the eye, then move to staph skin infections, and behind that we're working on a mix of respiratory staph, urinary tract infections and ultimately systemic staph infections where the phage would be injected."

The company had already raised $7.5 million and had completed 97% of its preclinical work. "Over the next few months, all of that data will be combined into an FDA application that we'll submit in late February [of 2002]," he said. "And hopefully we'll start trials in March."

Phage Therapeutics was also working on a treatment for tuberculosis and pseudomonas aeruginosa, which infects burn victims and cystic fibrosis patients.

He predicted that the day when doctors would be forced to diagnose the precise cause of each patient's infection was around the corner, driven by the rise of antibiotic resistance. "As doctors become more astute about the early identification of patients who are at risk of death," he said, "They are going to be more likely to demand earlier access to phages."

Chapter 10
Cows and Chickens

By December 2002, Sulakvelidze, originally feeling beaten down by his meeting with the FDA, was optimistic again. Intralytix had just signed a deal with Ecolab, a major distributor of everything from soap to food safety products. The deal brought an instant infusion of much-needed cash – a total of $1 million – to the company and promised a total of $4 million on a milestone basis. Ecolab would be responsible for the packaging and distribution of all of Intralytix's food-safety phages once they were approved by regulators. "To a degree, having a distributor is even more important than money, especially for a small company," said Sulakvelidze. "You have a product, you've been through the hurdles, the approval process. But how will you sell it? You have to make sure the packaging is right, that your product is being used effectively. A lot of things actually start after that point."

But perhaps Intralytix's greatest achievement lay in signing up a major industry partner: Perdue Farms. Perdue, a privately held company and the United States' second largest poultry producer after Tyson, had invested $1 million in the company and began experimenting with the use of phages to kill salmonella and listeria in its chickens, eggs and processing plants. "That was a big, big breakthrough for us," said Sulakvelidze. "Until then, I'd thought of phage therapy as a human thing, I thought this [deal with Perdue] is good, but … But then the more time went by, the more agricultural uses started to look interesting."

In that time, Intralytix also had contracted with a Baltimore company to do its large-scale production. So, within a year many things had fallen into place: the company had a client, a distributor and a manufacturer. All it needed was regulatory approval. And, as Intralytix prepared a new round of applications to the FDA, the Environmental Protection Agency and the United States Department of Agriculture – each of which would need to approve bacteriophages if they were to enter the U.S. food supply – it also embarked on a close collaboration with Perdue to produce specific phages for its needs and to generate good experimental data.

The deal with Perdue had started when one of Intralytix's founders, Gary Pasternack, mentioned the possibility of using phages in poultry to an acquaintance who was friends with Frank Perdue, the late chairman of the company. The acquaintance discussed bacteriophage with Frank one day over lunch, and the senior Perdue

A. Kuchment, *The Forgotten Cure: The Past and Future of Phage Therapy*,
DOI 10.1007/978-1-4614-0251-0_10, © Springer Science+Business Media, LLC 2012

referred Intralytix to his head of food safety, Keith Thompson. "They didn't know anything about phages," says Sulakvelidze. "But they said, 'Okay, come down and tell us about it.'" Sulakvelidze and Morris drove down to Salisbury, Maryland on the Delmarva Peninsula and made a simple presentation that explained what phages were and how Perdue could use them to their benefit. Thompson was there, in addition to Clay Silas, Perdue's chief microbiologist. They told Intralytix that they'd have to think about it. "Then, we got an invitation to present to a larger group," says Sulakvelidze. That time, they spoke before roughly 40 people from various divisions of the company, though Frank and his son Jim Perdue, the current CEO and President, were absent. "This time," says Sulakvelidze, "they said they wanted to try it, and we worked out the arrangements." Perdue now owns a small equity stake in Intralytix.

For Perdue, investing in phage therapy makes sense as poultry companies come under pressure to, on the one hand, ensure that their products are completely free of harmful bacteria and, on the other hand, to cut back on their use of antibiotics. In 2000, the Food and Drug Administration asked all poultry producers to stop the use of fluoroquinolones, the strongest of human antibiotics, for fear that they were contributing to the spread of antibiotic resistance. For years, meat producers have used antibiotics as growth-promoters to speed the time it takes their animals to get to market. In 2002, the largest American poultry companies, including Perdue, Tyson and Foster Farms, agreed to stop feeding antibiotics to chickens that weren't sick. In 2003 Perdue banned fluoroquinolones and, that same year, fast-food chains like McDonald's, Popeye's and Wendy's announced that they would refuse to buy chickens that had been treated with Baytril, a drug related to Cipro.

But at the same time, poultry producers need to continue reducing the levels of bacteria found in their products. A 2002 survey conducted in 24 metropolitan areas by Consumer Reports found that half of all chickens sold in supermarkets had been contaminated with bacteria: 42% had campylobacter; 12% had salmonella; 5% had both strains. More worrisome still was the statistic that 90% of the campylobacter strains and 34% of the salmonella strains showed resistance to one or more of nine antibiotics. Each year, 1.1 million Americans come down with food poisoning from eating undercooked chicken.

"[The idea of using bacteriophages] made a lot of sense to us, because it was a natural product, had no side-issues like toxicity, and didn't require huge pieces of equipment or tremendous cost," says Thompson, Perdue's former head of food safety, who no longer works for the company. "The poultry industry in general is trying to reduce the use of antibiotics, so new treatments like bacteriophage are becoming more appealing."

Perdue headquarters are located in a squat, beige stucco building a two hour drive from Wilmington, Delaware. Behind the reception desk hangs a large portrait of Arthur Perdue, Jim's grandfather and the founder of the company. By the desk is a small room, the Perdue "museum," that houses artifacts from company history, including photos of Dale Earnhart, the late NASCAR racer who was a poultry farmer and close friend of Jim's. There's also a mockup of a poster advertisement that has Frank Perdue posing as Albert Einstein, with electrified hair: "Developing a more tender, meaty breed of roaster takes a real genius."

Perdue's vice president for food safety, Bruce Stewart-Brown, a veterinarian who bares a striking resemblance to Jim Perdue, was overseeing the company's phage work. For the past few years, Perdue had been running extensive tests on bacteriophages against salmonella, by injecting them into eggs and spraying them onto chicks and fully grown chickens. So far, though the company declined to release exact statistics, it has found bacteriophages to be more effective than any other method it currently has for reducing salmonella, including vaccines and probiotics, though they had only tested bacteriophages on a small scale.

It may come as a surprise that Perdue even employs a microbiologist, but the poultry business runs on small profit margins that can be wiped out if even one bird gets sick. Because chicken houses are so crowded – holding about 24,000 birds apiece – epidemics can quickly spread out of control. So Clay Silas's lab is always busy analyzing microbiological samples from healthy birds, looking for signs of avian flu or any other illness that might burn through a flock. He can name all the common and not-so-common bacteria that are his birds' constant companions by strain, subspecies and serotype.

Perdue decided to start by testing phages on salmonella, because salmonella is the best-understood bacteria of the ones that infect chickens. Salmonella resides in chickens' intestinal tracts and is harmless to them – chickens rarely, if ever, fall ill from salmonella. But in humans it causes fever, chills, vomiting and diarrhea that linger for days. In the United States, salmonella infection results in 16,000 hospitalizations and 553 deaths per year, many of those not from undercooked poultry but from products made with uncooked eggs, like custards, pies and eggnog.

To combat salmonella, Perdue currently vaccinates its breeder birds – the hens that hatch the chickens that go to slaughter. The company also checks its chicken houses for excessive moisture, because they've learned that there's a relationship between moisture and the rate of salmonella colonization in their flocks. The drawback of the vaccines they now use is that, like bacteriophage, they're very specific to serotype. You can create antibodies to one type of salmonella, but they won't fend off another closely related type. But, with phages, Perdue and Intralytix had developed a cocktail with a good rate of crossover among salmonella serotypes. "We've tried a lot of different things that people bring us and that we think about, and it's really hard to get consistently good results," says Stewart-Brown. "But phage is one of those things where I would say, even if it didn't mean anything in the real world, so far it's better than other stuff we've tried." The major question that remained for Perdue was, once they start testing phages in larger chicken houses, how often will they have to update and remix their phage cocktails as a result of bacterial resistance?

Perdue has been testing its phages by spraying them on chicks and injecting them into eggs. On approximately the 17th day after an egg is laid it gets transferred from the setter, or incubator, to the hatchery. That's when poultry companies usually inject them with vaccines against various illnesses, like infectious bronchitis and Marek's disease, which can lead to paralysis and sudden death. As the eggs run along a conveyor belt, suction cups pick them up 200 at a time and transfer them to a machine that uses tiny needles to inject a small amount of medicine into the egg's air pocket, from where it seeps into the chick's mouth. "There's a very narrow window of time when you can do anything with an egg, between 17 and 19 days," says

Stewart-Brown. "If you went much earlier, the immune system of the chick wouldn't be working, and if you went any later, it would be hatched." Perdue has experimented with using phages at this same stage. Perhaps one day they will mix it with their other vaccines. After chickens hatch, they are vaccinated once more by going through a machine, 100 at a time, that sprays a fine mist over them. And phage have been used at this stage as well, along with a vaccine for Newcastle Disease, a severe respiratory infection.

The most frustrating aspect of working with Intralytix, says Thompson, has been waiting for the approach to win regulatory approval. "We have a great technology that can be of great benefit to everybody, especially to people in high-risk groups like children and pregnant women," says Thompson. "But the approval just takes so long. I think it's a very frustrating thing." Is Perdue likely to give up on Intralytix if the process takes too long? "They won't give up on it, I think," says Thompson. "But in a sense the ball is in Intralytix's court because they've got to as a company get the approvals. But Perdue won't give up on anything that's a benefit to the consumer. They'll spend a decade or more on it." Intralytix hopes so, too.

***

In January 2004 Ramachandran invited his Scientific Advisory Board to Bangalore for a major meeting. The company had recently hired a new CEO and was about to embark on a third round of financing where it hoped to raise $10 million – about five times the amount it had raised in each of the two previous rounds. These new funds, Ramachandran hoped, would last until the company brought its first product to market and began generating revenues. It was time to sit down and sharpen GangaGen's business strategy and streamline scientific priorities as executives prepared to woo a new group of investors.

GangaGen's scientific advisory board was the company's pride and its driving force. Few American scientists remained who had any expertise in bacteriophage biology, and it seemed that GangaGen had rounded up nearly all of them. Its scientific advisory board chairman was Gary Schoolnik from Stanford and members included Carl Merril, Sankar Adhya and Donald Court, a senior investigator at the NIH's National Cancer Institute who specializes in phage genetics. Ryland Young became GangaGen's executive director of research and for a couple of years split his time between GangaGen and Texas A&M. His expertise lies in the phage timing mechanism: how the viruses determine when to break out of their bacterial host. GangaGen's new CEO was David Martin, a physician and biochemist who was on the faculty of the University of California, San Franciso at the same time as Ram and later moved over to Genentech with Ramachandran. Since then, Martin had started up and sold a successful biotech company called EOS, specializing in cancer drug development. Though he does not have a background in phage biology, he has years of experience developing new drugs and leading them through clinical trials and negotiations with the FDA.

The scientists arrived on New Year's day and stayed at a spa that was a 45 minute drive from downtown Bangalore, down a dusty road crowded with roaming cows, bullock carts and brightly hand-painted pickup trucks. The first 2 days were spent on a bus tour of nearby temples and maharaja palaces from different eras of Indian history. As our minibus chugged past colorful villages, rice fields and carts stacked to the tipping point with sugar cane or hay, the conversation meandered from India's caste system to the nature of bacteriophages and the business of GangaGen. At one point, Martin recalled a conversation he'd had with a friend of his, a venture capitalist, about phage therapy. "It won't work," his friend told him. "Why not?" Martin had asked. "Arrowsmith's wife took phages and she died anyway," his friend replied incorrectly. When the tour guide mentioned that villagers used cow dung to sanitize the porches of their homes, the bus erupted in laughter: surely that was an example of bacteriophages at work.

By this time, GangaGen had raised another $2 million, a round that was completed on July 1, 2003 and led by Martin and Richard Spizzeri, a member of the board of directors. This, combined with ICF's initial investment, had allowed GangaGen to purchase and renovate two buildings in downtown Bangalore. The two buildings, modern and whitewashed, stand beside a highway on either side of an empty lot that was filled for a time with grazing cows. The main building was renovated and designed by an architect and fully equipped with three laboratories for only $200,000. The interior is clean and modern, with pale wood paneling and bulletin boards made from colorful Indian fabric. The staff, which now numbers 21, is made up of men and women in their early 20s, a generation that the Indian weekly magazine Outlook had dubbed "zippies" – young city or suburban residents between the ages of 15 and 25 "with a zip in the stride." "A growing slice of them (most Indians are still poor village-dwellers)," wrote New York Times columnist Thomas Friedman, "will be able to do your white-collar job as well as you for a fraction of the pay. Indian zippies are one reason outsourcing is becoming the hot issue in this year's U.S. presidential campaign."[1] In the lab, the women scientists wear saris or salwar kamizes topped with white lab coats.

Across the cow pasture is GangaGen's second building, which houses its future library, a conference room and the cafeteria, a spare modern room where a catering service lays out Tupperware containers of rice, naan, a few hot meat, lentil or chick-pea entrees and fresh fruit. The scientists serve themselves and do their own dishes.

Over the course of the weeklong meetings, the company reaffirmed its focus on targeting what it calls human pathogens in the food chain – in other words, bacteria like salmonella and e. coli that travel from animals like chickens and cattle into our food supply and cause outbreaks of disease. "To me, when you have a small company, the key is: focus, focus, focus," David Martin told me over beer and peanuts, sitting on the breezy veranda of a bar at the Angsana. "So we're really focused on E. coli 0,157."

E. coli 0,157 became a GangaGen priority shortly after Ramachandran had a chance run-in with an old friend of his from Berkeley. Michel Chretien, the brother

---

[1] Friedman, Thomas L., "Meet the Zippies," New York Times, Feb. 22, 2004, Section 4, pg. 11.

of former Canadian Prime Minister Jean Chretien, had heard about Ramachandran's fledgling company and thought phage therapy might be a perfect solution for a simmering crisis in Canada's food safety industry. In May 2000, *E. coli* 0,157 had killed 7 and sickened more than 600 residents of a small Ontario town. Walkerton, in the heart of a major beef-producing region, fell victim to a deadly contamination of its drinking water. *E. coli* 0,157, which lives in the intestinal tracts of cattle, had spread through their manure into the town's wells after a massive rain storm. When the local utility company's purification system malfunctioned, the bacteria seeped into the drinking water with fatal consequences. For 6 months, town residents were forced to rely on only bottled water and rinsed their hands in bleach after washing them – to prevent hand to mouth *E. coli* transmission. The outbreak prompted new nationwide legislation and stricter control over farms and groundwater supplies. "Michel said, 'This looks like the perfect, natural solution to this and he was very keen to look into it," said Ramachandran. Chretien, with additional funding from Canadian investors, set up a GangaGen subsidiary called GangaGen Life Sciences, based in Ottawa. Then he and Ramachandran brought on board Kishore Murthy, a young microbiologist who had previously worked at BioPhage, another Canadian company involved in phage work to run the research at GLS. "There are several different types of *E. coli* 0,157," Murthy told me at a dinner party that Ramachandran held in his Bangalore home. "We're looking into a cocktail that would kill all the variants. It works in the lab and, within a few months, we'll test it on cows and then will start looking for an industry partner to help us develop it." The attraction of starting by targeting *E. coli* in animals lies in the fact that the regulatory approval process would be far shorter and less expensive. With human trials, a company starts in the lab, then moves to animal trials and only then does several rounds of tests in humans. Scientists then have to follow the human subjects for several years to make sure there are no latent side-effects. "With food-chain applications," says Martin, "you just want to know: did you eliminate this human pathogen from the G. I. [gastro-intestinal] tract of the food animal so that the food doesn't get contaminated as the animal moves through the meat processors? So, you don't have to ask, 'What's the benefit a year or two after you start treating this patient with the disease?' You treat them a few weeks before they're slaughtered."

The GangaGen strategy was to put its *E. coli* product on the market within the next 2 years and then reinvest revenues from that into the race to develop a drug for human therapy, an undertaking that, Martin estimates, will end up costing about $100 million. "Because development is much more expensive than the research part [which comes earlier]," says Martin, "it's almost inevitable that one has to not only raise equity during that period of time, but you have to sell off many of your product ideas. You have to sell off your first born, your second born and hope you can hold on to the third born or afford to even *get to* your third born. Companies do partnerships with larger, deep pocketed pharmaceuticals or, these days, deep pocketed biotechnology companies in order to be able to survive and eventually develop a product."

That's why GangaGen's approach to human therapy must also be conservative. Instead of targeting the sexiest most dire pathogens, GangaGen would start with a preventative, topical product and slowly move toward one that would treat systemic illnesses like sepsis. While the company worked on *E. coli*, it is also investigating phage formulations to target Vancomycin- and Methicillin-resistant staph bacteria in hospitals. The treatment could be given to healthy hospital workers, perhaps in the form of a nasal spray, to prevent doctors and nurses from carrying and transmitting the bug to their patients.

And, though Ram's initial inspiration for pursuing phage therapy was to see it used as a tool to cure diseases in the developing world, that application, of necessity, had been placed on the backburner. "I would love to see a phage that we develop be used for diseases in the developing world," says Martin. But that's much easier said than done. Because there is no market for cholera or dysentery drugs in North America or Europe, the process of developing phages for those diseases would have to be funded entirely by a philanthropic agency – otherwise, GangaGen would not be able to recoup its investment. Far more likely, says Martin, is the prospect of a philanthropic group, like the Gates Foundation, stepping in and funding developing-world clinical trials for a drug that GangaGen developed for the First World, like one for disinfecting wounds and burns.

The conservative approach did not sit well with all the scientists on the SAB, many of whom came from academic or government backgrounds where private financing is not a concern. "There are people's lives at stake," said Merril one afternoon, as we boarded a bus en route to a tour of AstraZeneca's new headquarters. "I'm convinced that there are some illnesses for which only phages will work – and here we seem to be hearing a lot about paying back investors."

Sitting in Ramachandran's office at GangaGen, where two enlarged pages from d'Herelle's biography describing the Indian cholera study hang on the wall, I asked him how he felt about having to put off helping the developing world. "Unfortunately, nothing works just on a pure philanthropy," he said. "To take a phage through clinical trials is a very difficult challenge, and I believe that only a commercial company can do it – you have to have a strong incentive. Academic institutions are likely to get sidetracked – if something else came up, this would get dropped. But if it goes through a company, the company's survival depends on succeeding, and then it will happen."

His and Martin's hope was that, once any of the phage companies succeeds in conducting clinical trials according to modern practices, it will generate new interest in the field – especially from large pharmaceutical companies – and then there will be more money to put toward drug development in places like India. "If we are successful in the developed world," he said. "Then we'll be able to afford to provide this therapy to government institutions at a reasonable level. But I think this is the approach to take, it has to happen through a commercial entity that has some incentives built in to succeed. Just doing it on a purely non-profit basis I think is unlikely to work."

After the meetings, Martin and Ramachandran sounded confident that their company would be the one to succeed. Carlton, Sulakvelidze and Morris lacked a thorough understanding of phage biology, some members of the SAB suggested. "You need someone very talented to help get you past the FDA and USDA," one member said. "If neither the regulator nor the CEO has an understanding, the negotiations can bog down." Martin also suggested that his competitors were being naïve about intellectual property. In most industries other than medicine, it's enough to be first to market with a product and to claim a large share of consumers. But, says Martin, in biotechnology the cost of development is so high that a company has to be able to claim exclusivity to its product in order to maintain prices at a level at which they'll be able to pay back their investors and still make a profit. "Basically, if you don't have IP [intellectual property], you'll show your competitors how [to make your product] and they'll catch up with you pretty quickly," he says. "And even if they don't catch up with you they're gonna come up behind you and eat your lunch." Each successful biotech company is like a three legged stool, says Martin. One leg is intellectual property, one is an experienced management team that understands not only how to conduct research but that knows how to develop products, and the third is financing. "The company with the most solid three legs will come out on top – it will probably be the only one to come out on top," he says. "So, I think we've got the first two, and we're going to make certain that we put the third leg in place, and that's adequate funding."

Chapter 11
Approval, At Last

In August 2006, Intralytix scored a landmark victory: the FDA approved the use of phages in prepared food. This was the first time the agency had officially declared phages to be safe in any product under its purview. "A mixture of six bacteria-killing viruses can be safely sprayed on meat and poultry to combat common microbes that kill hundreds of people a year," announced the Associated Press, in a story that set a positive tone for the dozens of media reports that followed.

Intralytix's cocktail worked against listeria monocytogenes, a bacteria that sickens 2,500 Americans each year and kills 500; those most at risk are pregnant women, who are routinely told to steer clear of cold cuts, and those with weak immune systems. The FDA ruled that it was safe for companies to spray the product on luncheon meats, including bologna, turkey, chicken and roast beef, just before packaging.

"My jaw dropped when I looked at the number of news agencies that covered this story," said Sulakvelidze excitedly. In the days following the approval, he read hundreds of comments that readers left at the bottom of articles posted on the Internet, trying to ascertain how Americans felt about phages. Most consumers seemed to embrace the idea, but there were exceptions. "What if these viruses eat half my meat?" worried one reader on msnbc.com. "I pay too much already."

The Los Angeles Times, in an interview with an FDA spokesman, noted that the agency was fielding calls from consumers asking if products that had been sprayed with phages would be labeled as such. The answer was yes.

In the meantime, Intralytix greeted the news with tempered joy. Sulakvelidze bought his first bottle of Dom Perignon and uncorked it in the lab. But there were no major company-wide celebrations. "It was a long time since we saw any light, and we finally saw some light," said Sulakvelidze. "It took them almost a year, from the time they said they had no more technical questions, to issue this regulation, and that's a long time for a company of our size to be hanging out there."

One issue that complicated everything was that Ecolab, Intralytix's distribution partner, had bailed out on the company a few years before, and now Sulakvelidze

A. Kuchment, *The Forgotten Cure: The Past and Future of Phage Therapy*,
DOI 10.1007/978-1-4614-0251-0_11, © Springer Science+Business Media, LLC 2012

and his partners had to think through the details of packaging and distribution on their own. "We had a product but no money to manufacture it and market it, and there are a lot of little details: What label? What container? How do you ship it?" he said.

Intralytix was faced with a difficult choice: should they launch a major round of financing, or focus first on pushing out their product? "Our CEO, John Vazzana, decided to raise a little bit of money and get the rest through product sales," said Sulakvelidze.

***

GangaGen, in the meantime, shook up its management team and shed several members of its scientific advisory board. David Martin left the company following differences with Ramachandran, and eventually launched his own phage company in San Francisco. Carl Merril returned to his own research interests.

The company's priorities shifted as well. Ramachandran and his team had grown increasingly wary of submitting whole phages for FDA approval in humans. One observation that concerned him was his finding that some of his staph phages could kill only 75% of his staph bacteria, even though they had the ability to bind to all of them. After sequencing the resistant bacteria, he found that the cause of the resistance lay in a phenomenon known as superinfection immunity. If a bacterium becomes infected by a lysogenic phage – the type of phage that mixes its genes with bacterial genes but doesn't immediately burst out – that bacterial cell will become resistant to phages similar to the one inside it. While he could make sure that none of his phages were lysogenic, he couldn't control what kinds of bacteria his phages encountered in a patient's body. He decided it would be safer to come up with a drug that did not contain phage DNA. If Ramachandran could isolate a phage component with the ability to kill a bacterial cell as efficiently as a whole phage, not only might that product be more potent, but the FDA would approve it more readily.

After a few months, a GangaGen researcher isolated just such a component: a protein in the phage tail that's responsible for poking a hole in the bacterial wall. "When the phage binds to a specific receptor on a bacterial cell, there is a special enzyme that makes a tiny hole in the cell wall, and then DNA is injected into it," says Ramachandran. "But after the DNA is injected, the wall is resealed because the phage needs a viable bacterial host for at least several minutes to copy itself. But when we used the tails, all that happens is that a hole is made. There's no DNA going in to repair the hole, so it remains open, and eventually the cell dies because the membrane collapses." His current hope, he said, was to develop tails for every type of bacteria they wanted to target. If they failed, their fallback would be to seek approval for the lysine-deficient phage he had described in Bangalore.

Ramachandran was also excited about the company's progress on the *E coli* front. His researchers in Canada had tested their *E. coli* 0,157 phages in cattle in Alberta and found that they could rid the cows of the dangerous bacteria in 2–3 days.

GangaGen also showed that, once the phages kill the bacteria in the cows' intestines, the viruses do not remain in their system but are quickly flushed out.

***

For Intralytix, the months following the FDA approval were a blur of activity: marketing meetings, sales trips, efficacy tests to help satisfy customers' questions. "The cocktail continues to show wonderful efficacy: a 99% or so reduction in most foods we've tried," said Sulakvelidze. "It ranges from 96% up to 100%." But sales were lackluster. Perhaps it was a sign that consumers and food companies were more wary than they had first let on about adding viruses to their food. "It's just scary because everybody says stay away from viruses, and now we are eating them,"one expert told The Los Angeles Times after explaining that phages attack only bacteria, not humans. The overwhelming majority of the companies that did buy the product used it to clean surfaces in their processing plants rather than to spray it on their food.

The fall of 2007 brought a promising business opportunity that cheered up Sulakvelidze and his team. A British company contacted Intralytix saying that it was interested in buying a Scottish company that made smoked salmon. The problem: 52% of the smoked salmon market lies in the United States, which has a zero tolerance for listeria, a contaminant frequently found on smoked fish. Zero tolerance means that any batches of fish found to contain even trace amounts of listeria have to be recalled. The British company's CEO faxed a letter of intent to Intralytix, saying he would invest $25 million in Intralytix if Sulakvelidze could show that his product could clean the Scottish facility of 100% of its listeria. They also offered a $250,000 success fee on top of their investment. "We started packing," said Sulakvelidze.

He and Vazzana, his CEO, flew out to Scotland. At the plant, which had been sterilized the night before, he took samples of salmon and divided them into four groups. The first was left alone as a control, a second was treated with the FDA-approved phage cocktail, a third group was spiked with listeria and left untreated, and a fourth group was spiked with listeria and treated with phages. Sulakvelidze and Vazzana paced nervously for days in between testing sessions. "I don't remember how much I slept during that time," said Sulakvelidze. "There was nothing to do, but I didn't want to leave. It was a nerve wracking week." Finally, the results were ready. In the control group, 30% of the salmon samples were contaminated with listeria; in the spiked and untreated group, nearly all the salmon had high levels of the germ; but the treated groups had none. "The results were absolutely spectacular," he said.

Sandro and his CEO went out to a fancy dinner to celebrate, then briefly considered buying kilts to wear back to the office.

Unfortunately, the merriment was short lived. Soon after returning to Baltimore, Intralytix found out that the British company had gotten cold feet. The global economy was spiraling into recession, and the company had lost millions of dollars in investments. It paid Intralytix a success fee, but reneged on the $25 million investment.

"It was devastating for a number of reasons," says Sulakvelidze. "Number one, we didn't get funding that we thought we had. Number two, we sort of stopped raising money, because we thought: we have this deal, let's concentrate on the test. That was a setback, because we lost contacts we had before." Once the British company pulled out, other companies that had previously expressed interest in the Listeria cocktail, also grew nervous. "They thought, 'What do they know that we don't know?'" said Sandro. "It's hard to convince people once another investor pulls out. That was a major setback that we're still not fully recovered from," he said in August 2009.

***

That same summer, GangaGen was on the verge of starting pre-clinical trials on its phage tail product. Ramachandran's research team had further refined it and had come up with a clever way of making the protein even more potent. The company paired it with another protein, a lysostaphin, that binds to and kills bacteria in a similar manner. The two together worked beautifully. In rats infected with staph, the new drug, which GangaGen named StaphTame, wiped out nearly all the bacteria. Moreover, it would meet fewer regulatory hurdles than whole phages: StaphTame is known as a recombinant protein, a common class of drugs that includes blood-thinners and injectable insulin.

The StaphTame would first be used as a nasal spray. "The way we are approaching it is, 1 in 3 people carry staph in their nose, and it's harmless," says Ramachandran. "But then they go in for surgery or dialysis, and the staph gets into their body; 82% of the time, it's the patient's own nasal staph that gets in. The rest get it from health-care personnel." By helping to wipe out so-called nasal carriage, Ramachandran predicted his product would make a large dent in the number of people infected with MRSA and other lethal staph infections each year. GangaGen was preparing to launch Phase I and II trials in India and to petition the FDA to launch pre-clinical trials in the United States. If it received approval for StaphTame, GangaGen would test it to see if it could also help heal infected wounds.

The company was also in solid financial shape after a Japanese pharmaceutical company, Otsuka, made a large investment in GangaGen. One condition of that investment was that GangaGen should focus on human products exclusively; Ramachandran spun off GangaGen Life Sciences, the unit that focused on *E. coli* 0,157, and sold it to a Danish firm in 2008. "StaphTame is really the flagship of the company now," he said.

***

Tempering Intralyx's slow sales of its listeria product was a new interest on the part of the U.S. Army in developing phages for use in food safety products and to treat Iraq veterans with infected wounds. In early 2009, Intralytix signed a $1.5 million Army contract to develop a cocktail against *E. coli* 0,157, the deadly bacteria that had turned up recently in scallions from Taco Bell, bagged spinach and Nestle raw cookie dough. The cocktail would be sprayed on red meats, fruits and vegetables to help eradicate the germ.

Veterans of Afghanistan and Iraq, the Army felt, could also benefit from an additional weapon against a strain of bacteria that was showing increasing resistance to antibiotics in infected wounds: *Acinetobacter baumanii.* "American troops wounded in Iraq and brought back to military hospitals in the United States have unexpectedly high rates of infection with a drug-resistant type of bacteria," the New York Times reported in 2005. Intralytix was just beginning to isolate phages for the bugs, which are found in soil and water and which had also caused problems for Vietnam veterans. Once Acinetobacter invades a wound, it can spread easily to the bloodstream, bones and lungs.

"The next year is really critical for Intralytix," said Sulakvelidze. "We've been around for 10 years; we should just get it done or start looking for something else to do."

Chapter 12
In Treatment

The development of any new drug depends on the successful collaboration between physicians and researchers. In the case of phage therapy, surgeons like R. Nichol Smith, who saved actor Tom Mix; Alexander Tsulukidze, who pioneered the use of phages to treat surgical infections in the Soviet Union; and Gouram Gvasalia, who helped develop a powdered form of the drug to administer to his patients in Tbilisi, have all helped advance the medicine. In United States, however, it seemed phage therapy would be kept out of physicians' hands until the day Intralytix or Gangagen raised enough money to start clinical trials. In January 2006, Dr. Randall Wolcott of Lubbock, Texas found a better way.

Wolcott works in a nondescript building in downtown Lubbock, a dusty, sidewalk-free city in the Texas plains. He specializes in treating chronic wounds, an affliction that affects a booming segment of the population: diabetics and overweight, sedentary adults. The wounds form when a person's circulation is compromised, either by blood clots, which can form following joint replacement surgery, by diabetes, or by a hardening of the arteries called atherosclerosis. Invisible to the general public, chronic wounds are more lethal than many types of cancer and painful enough to hook the most stoic of adults on morphine.

Around 15% of diabetics develop leg ulcers, or breaks in the skin that are slow to heal. These wounds start small, caused by something as innocuous as an ill-fitting shoe, but, if infected, they can quickly spread over larger and larger areas of the foot and leg. Of all the diabetics who develop ulcers, more than 15% lose the affected foot or leg. Once a diabetic loses a leg, his risk of mortality quadruples. "[Diabetic amputees] have a very poor quality of life," says Wolcott. "Some 70% lose their other leg as well, and 80% are dead within 5 years."

Infected wounds are highly resistant to antibiotics, and not always because of the presence of superbugs. Once microbes penetrate a skin ulcer, they coat themselves in a protective layer that makes it difficult for antibiotics to reach them. This sugary coating, called biofilm, is a body-armor-like mesh that lets in nutrients for the bacteria but keeps out white blood cells, antibodies and, in most cases, drugs. "Everyone is so worried about [superbugs like] MRSA and VRSA," says Wolcott.

A. Kuchment, *The Forgotten Cure: The Past and Future of Phage Therapy*, DOI 10.1007/978-1-4614-0251-0_12, © Springer Science+Business Media, LLC 2012

"But the biofilm is, like, orders of magnitude of that problem, and nobody is talking about that."

Chronic wounds have a cure rate lower than that of breast, prostate, and colon cancer: around 50%. Thousands of these patients are quietly living out their lives across the United States, virtually immobilized and in unrelenting pain.

***

Wolcott calls his clinic "the only freestanding wound care center that is dedicated to helping each patient heal their wounds, no matter what it takes." A visit to his office reveals a remarkable array of tools that straddle old-world and cutting-edge medicine: molecular diagnostics, advanced dressings, bioengineered skin grafts, maggots, honey and the antacid Maalox. "We're just trying to use anything that's out there that can help us," he says. Maggots help clean the wound and stimulate circulation, certain kinds of honey inhibit the growth of bacteria ("Bees have a way of protecting their honey and keeping it bacteria-free," says Wolcott) and Maalox has shown some efficacy in breaking down biofilm.

Modern medicine's inability to conquer the chronic wound is what keeps Wolcott motivated and constantly on the hunt for new ideas. So, when an acquaintance first turned his attention to phages, the doctor threw himself into phage research. He soon got in touch with Betty Kutter and began learning all he could about the viruses.

Wolcott read phage studies dating back to the 1920s, met and spoke with phage researchers and biologists, attended Kutter's annual phage meeting at Washington's Evergreen University and even traveled to Tbilisi to observe Guram Gvasalia's and Zemphira Alavidze's work. What finally convinced him that phages were safe was a little-known fact: the only FDA-approved medical use of phages today is in AIDS patients. They are used to test patients' immune response.

Convinced bacteriophages could help his patients, Wolcott inquired about permission to use them in the United States. "I spent 3 months calling people, from my personal attorney, to the Texas Medical Association lawyers, to the Texas State Medical Board," he said. It turned out that, because phages are natural, Wolcott could use them in Texas without FDA approval, as long as he didn't stop standard treatments. "We did as much due diligence as we could do," says Wolcott, "and then we realized that ... well, we were losing some people we really shouldn't be losing."

***

Just as Wolcott was in the final stages of his phage research, one of his patients took a turn for the worse. Roy Brillon, 60, had struggled with blood clots in his legs since having knee replacement surgery 9 years before. The clots had damaged the valves in his veins, causing his lower legs to swell and his skin to break from the extra pressure. Brillon developed several infected ulcers on his legs. The largest was the size of the palm of his hand and a quarter inch deep. "I was in so much pain, it hurt even to lie in bed," he says. "Just a touch on the legs sent you into a nightmare."

Wolcott put Brillon on IV antibiotics and morphine. "I was supposed to take one [morphine pill] in the morning and one in the evening," says Brillon, a contractor with a round, friendly face and bright blue eyes. "But I was in so much pain, I started taking two in the morning, two in the afternoon and two at night. And then it even got worse, so I was taking three in the morning, three in the afternoon, three at night. Nothing was working." Wolcott's mission has been to end amputations for patients with chronic wounds. But here, even Wolcott was struggling not to lose hope.

"One morning, [Dr. Wolcott] got frustrated," said Brillon. "You could see it in his eyes. He said, 'Well, I'm seriously considering amputation.' My wife was sittin' there, and she went up through the ceiling. She said, 'No way, no way.'"

Wolcott turned around, walked out of the room, paced halfway down a long corridor, then turned around and came back into the examination room.

No, I'm not going to give up, said Wolcott.

That's my doc, said Brillon.

Wolcott decided to make Brillon his first phage patient.

***

Four months later, the first shipment of phages arrived from Tbilisi, Georgia. In her lab, Alavidze had matched the microbes in Brillon's wound, multidrug resistant *pseudomonas*, to her phages. At this point, Brillon was in the hospital, receiving round-the-clock care. Wolcott walked in and said, "I got the stuff from Russia." He removed Brillon's dressing, and put one drop of phage on each wound. "I've never seen a drop of anything that small," says Brillon. "I told him, 'We waited 4 months, and that's it?'" But, Brillon and Wolcott believe, the phages started working right away. "Within 36 h, the pain was cut in half," says Brillon. Within 6 weeks, a wound that had been open for years had almost completely closed up. And Wolcott noticed that the healing pattern, with phages, was different from what he normally sees. Most wounds close from the outside in: the circumference shrinks and the skin gradually knits together. But Brillon's wound also grew new skin in the center, called an epithelial island, that spread outward. To the amazement of Brillon and Wolcott, who are both religious, the new skin formed the shape of a cross in the center of his healing ulcer. They took pictures to document it.

After phages seemingly rescued Brillon's legs, Wolcott thought he might have found a miracle drug. He began purchasing phages from Intralytix and sequencing each one to make sure it was free of genes for lysogeny and pathogenicity. He hired a staff to work with his phages, including a visiting student from Tbilisi and a young lab tech named Dan Rhoads, whom he had met at Kutter's annual phage meeting.

Soon, phages grew into a regular part of Wolcott's clinical practice. He added them to his debriders – machines that doctors use to cleanse and remove dead tissue from wounds. Each debrider has a wand that sends ultrasound waves, through a water-based solution, into patients' wounds. To that solution, Wolcott added phages, and they worked well. While he found that only 1–5% of patients had the same miraculous response as Brillon, 30% showed some clear benefit. "In these types of wounds, where antibiotics just aren't that effective, that's really good," he said.

He felt he had to find a way to make phages a part of standard U.S. medical practice, so he petitioned the FDA to let him perform a physician-initiated Phase I clinical trial, to prove phages are safe for use in wounds. When Intralytix couldn't put up the money to fund the trial, Wolcott paid for it himself: $500,000 over the course of 3 years.

In June 2009, his study was published in the Journal of Wound Care: out of a group of 39 treated patients – near the 40 person maximum allowed in a Phase I trial – none had suffered side effects from the phages. The FDA gave Wolcott a green light to proceed with Phase II, which would test for efficacy.

Here, things ground to a halt. As Wolcott and Intralytix were launching their safety trial, the FDA explained that, to get through Phase II, the company would have to put each phage through a separate trial. The cost of that, said Wolcott, would be astronomical. To treat all the germs his patients come in with would require a collection of 100–1,000 phages. (The Eliava Institute's collection is made up of at least 2,000 phages.) Putting them each through a clinical trial would cost $100 million at a minimum. "The FDA should approve the process, not the individual phages," he says. Flu vaccines, for example, do not undergo new trials each time they're reconstituted, because the FDA has approved the basic process by which they are assembled and grown up in the lab. "They are denying patients a very rational treatment," he says.

Epilogue

Phage therapy has come a long way since I first started working on "The Forgotten Cure" in 2001. Back then, there were only a handful of researchers interested in phages as a potential alternative to antibiotics, and their efforts were largely ignored or derided. "There are two kinds of drugs: those you would give your mother, and those you would give your mother-in-law," a pharmaceutical executive told me in an interview for Newsweek International that year. "Phages are what you would give your mother-in-law." Alexander Sulakvelidze of Intralytix recalls that funding agencies used to throw his applications in the trash instead of taking the time to review and score them. "A lot of people told us, 'There is no way you can ever get this approved in the U.S.'" he says. "And it was against all odds that we kept pushing."

The landscape of pathogenic bacteria has evolved since 2001 as well. At the time, many infectious disease experts, including Intralytix co-founder J. Glenn Morris, were preoccupied with VRE, the superbug that colonizes the intestinal tract and seeps into the bloodstream when people become very ill. Ten years later, VRE remains confined largely to hospitals, and a newer antibiotic called Linezolid has managed to keep it in check. "We haven't by any means solved the problem, but we've bought ourselves some time," says Morris, who also serves as director of the Emerging Pathogens Institute at the University of Florida. His chief sources of concern today are MRSA, a strain of the common skin bacteria Staphylococcus aureus that is resistant to the antibiotic methicillin; and a class of bacteria called Gram-negatives, some of which have become resistant to all antibiotics. "What makes MRSA so scary is that it is spreading rapidly through the community," says Morris. That means healthy individuals are as susceptible as those in the hospital, and experts can exercise little control over the superbug's spread.

The case of Gram negative bacteria is at least as jarring. This class of germs, which includes E. coli and Acinetobacter baumannii, is defined by its double cell membrane and its response to a specific staining process developed by Danish physician Hans Christian Gram. The double membrane helps insulate the bacteria from antibiotics, but phages have been breaching this extra barrier for millennia. In 2009, infectious disease experts began seeing new strains of Gram-negative bacteria that were resistant even to carbapenems, a class of antibiotics that are considered a last resort. This has forced doctors, in the worst cases, to rely on older, previously discarded drugs that can cause kidney damage. "Clinical microbiologists increasingly agree that multidrug-resistant Gram-negative bacteria pose the greatest risk to public health," wrote the authors of a September, 2010 study in the journal The Lancet Infectious Diseases. "Not only is the increase in resistance of Gram-negative bacteria faster than in Gram-positive bacteria, but also there are fewer new and developmental antibiotics active against Gram-negative bacteria."[1] What has troubled clinicians most of all is the fact that these newly discovered resistance genes are easily passed among bacteria and that they have already affected E. coli, which is the most common cause of a very common ailment: urinary tract infections. "The consequences will be serious if family doctors have to treat infections caused by these multiresistant bacteria on a daily basis," noted Johann Pitout of the University of Calgary in an editorial that accompanied the Lancet Infectious Diseases study.[2]

Bacteriophage researchers are working to address these and many other challenges. In June, 2009 the British company Biocontrol Limited delivered what the field has long lacked: conclusive evidence that phage therapy works. In the first successful Phase II clinical study of the treatment, researchers showed that a single drop of phage cocktail was enough to cure three out of 12 patients with chronic, antibiotic-resistant ear infections caused by the Gram-negative germ pseudomonas aeruginosa; all but one patient in the treated group improved. In an untreated control group, none were cured and three deteriorated. In a larger Phase III trial, Biocontrol Limited, which was acquired by Seattle, Wash.-based Targeted Genetics Corporation in early 2011, will look into whether follow-up treatment with phages would result in a higher cure rate.[3]

GangaGen's Ram Ramachandran is moving ahead with a drug that could prevent MRSA infections. He held a successful meeting with the Food and Drug administration in the spring of 2010 regarding StaphTame, his anti-MRSA nasal spray. The agency approved, in principal, his plans for a combined safety and efficacy trial of the phage-derived drug, and he was hoping to launch the testing in 2011.

Richard Honour, the former partner of Caisey Harlingten, the investor who helped ignite a renewed global interest in phage therapy in the 1990s, is also working on MRSA. Honour is president and CEO of Viridax, based in Boca Raton, Fla., which specializes in phages that have been genetically modified to treat specific strains of MRSA known to cause Ventilator-Associated Pneumonia. "It's not a big market, but it's an important one," says Honour. Fifty percent of elderly patients who go on a ventilator and contract MRSA pass away, he says. The company has

[1] Woodford, Neil, et al. "Emergence of a new antibiotic resistant mechanism in India, Pakistan and the UK: a molecular, biological and epidemiological study." The Lancet Infectious Diseases, Volume 10, Issue 9, pages 597–602. Sept. 2010.

[2] Pitout, Johann D D, "The latest threat in the war on antimicrobial resistance." The Lancet Infectious Diseases, Volume 10, September 2010.

[3] Harper, D.R. et al. "A controlled clinical trial of a therapeutic bacteriophage preparation in chronic otitis due to antibiotic-resistant *Pseudomonas aeruginosa*; a preliminary report of efficacy." Clinical Otolaryngology, 2009 **34**, 349–357.

changed its business strategy several times due to the tough investment climate in the United States and is considering incorporating Viridax in Switzerland instead. "We would then have an easier route to financing over there," he says. Viridax's first product is at least 5 years away from market.

For Intralytix, sales of its listeria product for food safety are up after a disappointing start. As of 2010, the company was selling the product, known as ListShield, regularly to a half-dozen seafood companies. At a meeting in New York City, Sulakveldize showed off a new brochure for his anti-listeria and anti-E. coli food products, noting that they had been certified organic, Kosher and Halal. One new ally on the food safety front has been Frank Costanzo, plant manager for the Williamsburg, Brooklyn-based Service Smoked Fish. Costanzo has tested ListShield on his equipment as well as by spraying it directly on smoked salmon – though is not yet selling any phage-sprayed salmon to the public – and says it is the only product that has consistently been able to wipe out all traces of listeria from his seafood. Costanzo explains that state and FDA inspectors will recall any batch of fish with even a trace amount of the organism. Yet the microbes are widely distributed in nature, making them virtually impossible to eradicate; the germs, which can cause gastrointestinal symptoms in adults and meningitis in newborns, are the reason that pregnant women are advised to steer clear of smoked fish, soft cheeses and ready-to-eat meats. Costanzo says the phages are an attractive alternative to the chemicals in which some companies dip their fish to pass inspection.

The Army has also funded a novel Intralytix product aimed at protecting soldiers and civilians from some forms of travelers' diarrhea. The idea is to use phages as probiotics by swallowing them in the form of pills or mixing them into yogurt. Though phages are not typically thought of as probiotic organisms, their function is similar: they naturally populate the gut and can inhibit the growth of pathogenic bacteria. He says phages fit the definition of probiotics as set out by the United Nations' Food and Agriculture Organization and the World Health Organization.

Two entrepreneurs mentioned earlier in the book have left the phage therapy field. Caisey Harlingten has been out of the business since the early 2000s. After splitting with Richard Honour, he moved to England and co-founded a new company called Regma Bio-Technologies, Inc. with Soviet defector and bioweapons expert Vladimir Pasechnik. Pasechnik had filed patents for the use of phages against tuberculosis, which were to form the heart of the company. Unfortunately, he died unexpectedly in 2001, and the company's business ground to a halt. "We tried to carry on after Vlad died, but it became apparent that it wouldn't work," said Harlingten in 2010. "The whole project became too costly and too time consuming, and we realized that we could carry on for years before we were lucky enough to get our product functional in an animal model." In the end, Harlingten sold the remains of Regma Biotechnologies to a gold mining company. Today, he invests in solar cells and splits his time between Canada, the UK and France. Does he regret investing in phage therapy? No, says Harlingten. He learned that there were certain ailments for which phages could work; he traveled well and learned interesting things. And, he notes, he made back all of his money, times two.

Richard Carlton, who worked with Carl Merril on one of the first phage therapy companies in the United States is, like Harlingten, now out of the business. He sold the food safety arm of Exponential Biotherapies to a group of Dutch investors who formed EBI Food Safety, a competitor to Intralytix. According to its Web site, EBI Food Safety markets its own anti-listeria product, Listex, for use in cheese, meat, fish and poultry processing plants.

***

One fact about phage therapy that has not changed significantly in the last decade is that there still have been no large-scale clinical trials showing the treatment is effective. This results from a frustrating catch-22: to initiate trials, companies need to raise money, but investors and pharmaceutical companies prefer to invest in ideas that have data to back them up. Companies like Intralytix, GangaGen and Biocontrol Limited have persisted, and they may finally be on the verge of gaining some more definitive answers.

Until then, patients with chronic infections for whom antibiotics have failed will continue to be drawn to Tbilisi for treatment. In recent years, the Eliava Institute has taken steps to make its services more widely available and is currently planning to open offices in Queens, N.Y. and in Nurnberg, Germany that will screen patients for possible medical trips to Tbilisi.

The San Ramon, Calif.-based company Phage International now contracts with some Eliava scientists to bring patients to Georgia for treatment. Founded by entrepreneur Chris Smith in 2004, Phage International has helped around 100 patients as of 2010, and says Smith, nearly all of them have recovered thanks to careful pre-screening. He first encountered phage therapy in 2002, while sitting in the waiting room of the U.S. embassy in Tbilisi waiting for a flight back to the States. The Iraq war had broken out while he and his wife were in Georgia adopting their daughter. In an embassy publication, he came across an interview with Betty Kutter, the Evergreen College phage biologist who has served as a liaison between the Eliava Institute and the West. "We all knew about the problem of antibiotic resistance," he recalls. "And this seemed like an opportunity." He contacted Kutter, spent time at the Eliava Institute and contracted with Zemphira Alavidze and a group of Tbilisi physicians to identify patients who could benefit from phage therapy. Alavidze and Guram Gvasalia are on the board; Randy Wolcott, the Lubbock, Texas-based wound care expert is an investor. In the last 6 years, Smith has arranged trips for patients from the U.S., Canada, Australia, New Zealand and parts of Africa; the conditions range from chronic prostatitis to sinusitis to urinary tract infections. His best track record, he says, is with prostatitis. "Our medical system cannot deal with it at all," he says, "But we get fantastic results." The Eliava scientists, he points out, administer not only phages but a mix of treatments, including antibiotics. "The bacteria tend to become more sensitive to antibiotics during treatment."[4]

---

[4] This is a claim made repeatedly by Eliava scientists and backed up by Western literature. See Chapter III for a Lancet paper from the 1940s about how phages and penicillin can work together.

His first case was, perhaps, his most dramatic. Laura Roberts, a single mother from Fort Worth, Texas, had suffered for more than a decade from a chronic sinus infection brought on by severe allergies. The combination spiraled out of control: the allergies lead to asthma, and the infection grew resistant to antibiotics. "I just kept getting worse and worse," says Roberts, now 57. She developed a host of other medical problems: bone loss from the steroids she was taking for asthma; hip replacement surgery for the bone loss; fibromyalgia, migraines, fatigue, and weight loss. Her otolaryngologist, Dr. Natalie Roberge, of Texas Ear, Nose & Throat, had Roberts on and off IV antibiotics as well as IV immunoglobulins to boost her immune system. Roberts was seeing 12 specialists, including infectious disease experts, a rheumatologist and a neurologist. "She just wouldn't respond to anything," said Roberge.

Finally, in September, 2005, Roberge sent Roberts to the Mayo Clinic in Rochester, Minn. Roberts spent 9 days there, working with some of the country's top experts on her condition: allergic fungal sinusitis, an allergy where the body overreacts to molds in the air. In her case, the constant swelling in her nasal passages had led to the growth of staph aureus bacteria, which at first were sensitive to antibiotics but later bred resistance. The doctors at the Mayo Clinic threw up their hands. There were no antibiotics left to try. "They said they didn't expect me to live past the end of 2005," she says.

That's when Roberts remembered a program she had seen on television that mentioned the Eliava Institute – the same episode of "48 h" that Saharra Bledsoe had watched in Fort Wayne, Indiana. Back home in her bed, where she had been confined for months before her trip to Rochester, she called her younger brother, Andrew Silva, and asked him to investigate. In the years between Fred Bledsoe and Laura Roberts, Phage International had sprung up; now communication with Tbilisi was easier, if more expensive (individual treatments cost $5,000 to $15,000, not including airfare and hotel). Dr. Roberge was skeptical about her patient's plans but gave her consent. "I've never been somebody who's big into alternative medicine, but at that point, she was so sick, and we were making no progress," she says. Roberge had bacterial samples prepared and sent to Tbilisi, where Alavidze quickly found matching phages, and Roberts and her brother quickly booked a flight out.

In Tbilisi, Alavidze and a team of doctors treated Roberts in a new, modern clinic. Roberts and her brother went there every day for 3 weeks for treatment with a variety of medicines, including phages. Roberts describes a rigorous morning regimen at the clinic: workers would flush out her nasal and ear passages with saline, then apply electrical stimulation, known as TENS, to her sinuses to break up the mucous; then, using long thin instruments, they would apply gauze soaked in a honey extract called camylin to her nasal passages. Honey is believed to have antibiotic and anti-inflammatory properties. Finally, the doctors would drip phages into her sinuses, and "blow" powdered dried phages into her ears. At night, in her hotel, Roberts would put a few more drops of phages into her sinuses.

She says she started feeling better almost immediately. On her first night following treatment, she kept her dinner down for the first time in recent memory. The next day, she felt hungry for the first time in years. A few days later, she shed the socks and gloves she wore even indoors for warmth. "By the third week, we were doing things like going to the museum or going site-seeing after her treatments," said Silva. "We couldn't believe it. Toward the end, it felt like we were on a little vacation."

Roberts had arrived with a cane and walker but left without them. Back in Fort Worth, her ENT, Dr. Roberge couldn't believe her eyes. "It was remarkable how much better she looked," says Roberge. Roberts used to come into the office every week to have her sinuses and Eustachian tubes drained. But now, everything was dry. Because Roberts was treated with many different things, it's hard to know if it was really phages that cured her. But, to Dr. Roberge, the evidence is pretty clear. "It was either phages, or it was a miracle," she says.

Five years later, Roberts has returned to Tbilisi twice to treat two new infections. In total, she says, she has spent $12,000 from her savings on the treatments, not including hotel and airfare. But, to her, it was worth it. She's resumed all her previous activities, including taking care of her daughter, who is now in college.

"We do know that in single cases, phages work, but there are still so many things that science needs to research before phages can be used widely," says Lasha Gogokhia, who trained under Guram Gvasalia in Tbilisi and is now a postdoctoral fellow in pathology at the University of Utah in Salt Lake City. Among the things scientists need to learn: how do phages reach bacteria? How can researchers make them work more efficiently and effectively? Why are there cases where phage therapy fails? Until researchers find the money and the resources to uncover those answers, phage therapy's promise will remain unfulfilled.

About the Author

Anna Kuchment is an editor at Scientific American magazine. She first wrote about bacteriophages for Newsweek magazine, where she was a staff writer covering international affairs, science, health and the environment. Her investigation into phage therapy took her around the world to India, Poland, France and the former Soviet Union. Born in Moscow, she is a graduate of the Columbia University Graduate School of Journalism and lives in New York City. www.annakuchment.com

Index

CPSIA information can be obtained at www.ICGtesting.com
Printed in the USA
LVOW081722141112

307333LV00002B/1/P